How to Read and
Do Proofs

How to Read and Do Proofs

An Introduction to Mathematical Thought Processes

Third Edition

Daniel Solow
Department of Operations
Weatherhead School of Management
Case Western Reserve University
Cleveland, OH 44106
e-mail: dxs8@po.cwru.edu
web: http://weatherhead.cwru.edu/solow/

John Wiley & Sons, Inc.

New York / Chichester / Weinheim / Brisbane / Singapore / Toronto

ACQUISITIONS EDITOR	Kimberly Murphy
MARKETING MANAGER	Julie Z. Lindstrom
SENIOR PRODUCTION EDITOR	Ken Santor
SENIOR DESIGNER	Karin Kincheloe
COVER DESIGNER	Carol Grobe

This book was set in LaTeX by the author and printed and bound by Courier Westford. The cover was printed by Phoenix Color Corporation.

This book is printed on acid free paper.

ISBN 0-471-40647-3   (pbk.)

Printed in the United States of America

10 9 8 7 6 5 4 3 2

To my late father, Anatole A. Solow,
and to my mother, Ruth Solow

Contents

Foreword   *ix*

Preface to the Student   *xi*

Preface to the Instructor   *xiii*

Acknowledgments   *xvii*

1   Chapter 1: The Truth of It All   *1*

2   Chapter 2: The Forward-Backward Method   *9*

3   Chapter 3: On Definitions and Mathematical Terminology *23*

4   Chapter 4: Quantifiers I: The Construction Method   *35*

5   Chapter 5: Quantifiers II: The Choose Method   *45*

6   Chapter 6: Quantifiers III: Specialization   *59*

7   Chapter 7: Quantifiers IV: Nested Quantifiers   *69*

8   Chapter 8: The Contradiction Method   79

9   Chapter 9: The Contrapositive Method   91

10 Chapter 10: Nots of Nots Lead to Knots   99

11 Chapter 11: Uniqueness Methods and Induction   105

12 Chapter 12: Either/Or and Max/Min Methods   121

13 Chapter 13: Summary   133

Appendix A Putting It All Together: Part I   145

Appendix B Putting It All Together: Part II   153

Solutions to Selected Exercises   161

Glossary   193

References   199

Index   201

Foreword

In a related article, "Teaching Mathematics with Proof Techniques," the author has written, "The inability to communicate proofs in an understandable manner has plagued students and teachers in all branches of mathematics." All of those who have had the experience of teaching mathematics and most of those who have had the experience of trying to learn it must surely agree that acquiring an understanding of what constitutes a sound mathematical proof is a major stumbling block for the student. Many students attempt to circumvent the obstacle by avoiding it—trusting to the indulgence of the examiner not to include any proofs on the test. This collusion between student and teacher may avoid some of the unpleasant consequences—for both student and teacher—of the student's lack of mastery, but it does not alter the fact that a key element in mathematics, arguably its most characteristic feature, has not entered the student's repertoire.

Dr. Solow believes that it is possible to teach the student to understand the nature of a proof by systematizing it. He argues his case cogently with a wealth of detail and examples in this book, and I do not doubt that his ideas deserve attention, discussion, and, above all, experimentation. One of his principal aims is to teach the student how to read the proofs offered in textbooks. These proofs are, to be sure, not presented in a systematic form. Thus, much attention is paid—particularly in the two appendices—to showing the reader how to recognize the standard ingredients of a mathematical argument in an informal presentation of a proof.

There is a valid analogy here with the role of the traditional algorithms in elementary arithmetic. It is important to acquire familiarity with them and to understand why they work and to what problems they could, in principle, be applied. But once all of this has been learned, one would not slavishly execute the algorithms in real-life situations (even in the absence of a calculator!). So, the author contends, it is with proofs. Understand and analyze their structure—and then you will be able to read and understand the more informal versions you will find in textbooks—and finally you will be able to create your own proofs. Dr. Solow is not claiming that mathematicians create their proofs by consciously and deliberately applying the "forward-backward method"; he is suggesting that we have a far better chance of teaching an appreciation of proofs by systematizing them, than by our present, rather haphazard, procedure based on the hope that the students can learn this difficult art by osmosis.

One must agree with Dr. Solow that, in this country, students begin to grapple with the ideas of mathematical proof far too late in their student careers—the appropriate stage to be initiated into these ideas is, in the judgment of many, no later than eighth grade. However, it would be wrong for university and college teachers merely to excuse their own failures by a comforting reference to defects in the student's precollege education.

Today, mathematics is generally recognized as a subject of fundamental importance because of its ubiquitous role in contemporary life. To be used effectively, its methods must be understood properly—otherwise we cast ourselves in the roles of (inefficient) robots when we try to use mathematics, and we place undue strain on our naturally imperfect memories. Dr. Solow has given much thought to the question of how an understanding of a mathematical proof can be acquired. Most students today do not acquire that understanding, and Dr. Solow's plan to remedy this very unsatisfactory situation deserves a fair trial.

PETER HILTON

Distinguished Professor of Mathematics
State University of New York
Binghamton, NY

Preface to the Student

After finishing my undergraduate degree, I began to wonder why learning theoretical mathematics had been so difficult. As I progressed through my graduate work, I realized that mathematics possessed many of the aspects of a game—a game in which the rules had been partially concealed. Imagine trying to play chess before you know how all of the pieces move! It is no wonder that so many students have had trouble with abstract mathematics.

This book describes some of the rules by which the game of theoretical mathematics is played. It has been my experience that virtually anyone who is motivated and who has a knowledge of high school mathematics can learn these rules. Doing so greatly reduces the time (and frustration) involved in learning abstract mathematics. I hope this book serves that purpose for you.

To play chess, you must first learn how the individual pieces move. Only after these rules have entered your subconscious can your mind turn its full attention to the more creative issues of strategy, tactics, and the like. So it appears to be with mathematics. Hard work is required in the beginning to learn the fundamental rules presented in this book. In fact, your goal should be to absorb this material so that it becomes second nature to you. Then you will find that your mind can focus on the creative aspects of mathematics. These rules are no substitute for creativity, and this book is not meant to teach creativity. However, I do believe that the ideas presented here provide you with the tools needed to express your creativity. Equally important is the fact that these tools enable you to understand and appreciate the creativity of others. To that end, much emphasis is placed on teaching you how to read "condensed" proofs as they are typically presented in textbooks, journal articles, and other mathematical literature. Knowing how to read and understand such proofs enables you to assimilate the material in any advanced mathematics course for which you have the appropriate prerequisite background. In fact, knowing how to read and understand condensed proofs gives you the ability to learn virtually any mathematical subject on your own, with enough time and effort.

You are about to learn a key part of the mathematical thought process. As you study the material and solve problems, be conscious of your own thought processes. Ask questions and seek answers. Remember, the only unintelligent question is the one that goes unasked.

DANIEL SOLOW

Department of Operations
Weatherhead School of Management
Case Western Reserve University
Cleveland, OH

Preface to the Instructor

The Objective of This Book

The inability to communicate proofs in an understandable manner has plagued students and teachers in all branches of mathematics. The result has been frustrated students, frustrated teachers, and, oftentimes, a watered-down course to enable the students to follow at least some of the material, or a test that protects students from the consequences of this deficiency in their mathematical understanding.

One might conjecture that most students simply cannot understand abstract mathematics, but my experience indicates otherwise. What seems to have been lacking is a proper method for explaining theoretical mathematics. In this book I have developed a method for communicating proofs—a common language that professors can teach and students can understand. In essence, this book categorizes, identifies, and explains (at the student's level) the various techniques that are used repeatedly in virtually all proofs.

Once the students understand the techniques, it is then possible to explain any proof as a sequence of applications of these techniques. In fact, it is advisable to do so because the process reinforces what the students have learned in the book.

Explaining a proof in terms of its component techniques is not difficult, as is illustrated in the examples of this book. Before each "condensed" proof is an analysis explaining the methodology, thought processes, and techniques that are used. Teaching proofs in this manner requires nothing more than preceding each step of the proof with an indication of which technique is about to be used, and why.

When discussing a proof in class, I actively involve the students by soliciting their help in choosing the techniques and designing the proof. I have been pleasantly surprised by the quality of their comments and questions. It has been my experience that once students become comfortable with the proof techniques, their minds tend to address the more important issues of mathematics, such as why a proof is done in a particular way and why the piece of mathematics is important in the first place. This book is not meant to teach creativity, but I do believe that learning the techniques presented here frees the student's mind to focus on the creative aspects. I have also found that, by using this approach, it is possible to teach subsequent mathematical material at a more sophisticated level without losing the students.

In any event, the message is clear. I am suggesting that there are many benefits to be gained by teaching mathematical thought processes in addition to mathematical material. This book is designed to be a major step in the right direction by making abstract mathematics understandable and enjoyable to the students and by providing you with a method for communicating with them.

Why a Third Edition?

After teaching the material in this book for more than 20 years to undergraduate mathematics majors and to graduate students in operations research and computer science, I have experienced how effective this systematic method is for learning proofs. At the same time I have learned how to improve the approach significantly. This third edition incorporates the following improvements, of which the first two are the most significant, as well as many of the comments I received from various colleagues via a questionnaire distributed by the publisher, John Wiley & Sons.

- The inclusion, in practically every chapter, of new material on how to read and understand proofs as they are typically presented in class lectures, textbooks, and other mathematical literature. The goal is to provide sufficient examples (and exercises) to give students the ability to learn mathematics on their own.

- A complete revision of the exercises, including many new ones on reading condensed proofs in which the students are asked to explain those proofs in terms of the techniques they have learned in this book. Some of the exercises, as marked with a B, have answers in the back of the book; some exercises, as marked with a W, have answers on the web at http://www.wiley.com/college/solow/; and the rest, whose answers are provided only in the accompanying Solutions Manual for the instructor.

- A reorganization that groups the special quantifier techniques of induction and uniqueness together in Chapter 11.

- A glossary that includes all definitions of mathematical terms for easy access and reference.

- A bibliography of related books on proofs and general mathematical reasoning.

- Improved explanations together with some new examples.

Although these changes seem to make it even easier for students to understand proofs, I have still found no substitute for actively teaching the material in class instead of having the students read the material on their own. This active interaction has proved eminently beneficial to both student and teacher, in my case.

DANIEL SOLOW

Department of Operations
Weatherhead School of Management
Case Western Reserve University
Cleveland, OH

Acknowledgments

Acknowledgments for the First Edition

For helping to get this work known in the mathematics community, my deepest gratitude goes to Peter Hilton, an outstanding mathematician and educator. I also thank Paul Halmos, whose timely recognition and support greatly facilitated the dissemination of the knowledge of the existence of this book and teaching method. I am also grateful for discussions with Gail Young and George Polya.

Regarding the preparation of the manuscript for the first edition, no single person had more constructive comments than Tom Butts. He not only contributed to the mathematical content but also corrected many of the grammatical and stylistic mistakes in a preliminary version. I suppose that I should thank his mother for being an English teacher. I also acknowledge Charles Wells for reading and commenting on the first handwritten draft and for encouraging me to pursue the project further. Many other people made substantive suggestions, including Alan Schoenfeld, Samuel Goldberg, and Ellen Stenson.

Of all the people involved in this project, none deserves more credit than my students. It is because of their voluntary efforts that the first edition of this book was prepared in such a short time. Thanks especially to John Democko for acting in the capacity of senior editor while concurrently trying to complete his Ph.D. program. Also, I appreciate the help that I received from Michael Dreiling and Robert Wenig in data basing the text and in preparing the exercises. Michael worked on this project almost as long as I did. A special word of thanks goes to Greg Madey for coordinating the second rewriting of the document, for adding useful comments and, in general, for keeping me very organized. His responsibilities were subsequently assumed by Robin Symes.

In addition, I am grateful to Ravi Kumar for the long hours he spent on the computer preparing the final version of the manuscript, to Betty Tracy and Martha Bognar for their professional and flawless typing assistance, and to Virginia Benade for her technical editing. I am also indebted to my class of 1981 for preparing the solutions to the exercises.

I thank the following professors for refereeing the original manuscript and for recommending its publication: Alan Tucker, David Singer, Howard Anton, and Ivan Niven.

Acknowledgments for the Second Edition

Most of the credit for the improvements in the second edition goes to the students who have taught me so much about learning mathematics. I am also grateful for the many comments and suggestions I have received from colleagues over the years and, more recently, on a questionnaire circulated by John Wiley.

On the technical end, I thank Dawnn Strasser for her most professional job of typing the preliminary version of the manuscript. I also am grateful to the Weatherhead School of Management at Case Western Reserve University for the use of their excellent word-processing facilities.

Finally, I am grateful to my wife, Audrey, for her help in proofreading and for her patience during yet another of my projects.

Acknowledgments for the Third Edition

The third edition was typeset using LATEXwith thanks to Donald Knuth and also Amy Hendrickson, whose macro package was a delight to use. Serge Karalli spent many hours helping me revise old exercises and develop new ones. Tom Engle created professional electronic drawings of the figures. I also thank John Wiley & Sons for providing me with the support I needed to accomplish this revision. It was also a pleasure to work with my editor, Kimberly Murphy and her staff, especially Ken Santor.

D. S.

1

The Truth of It All

The objective of mathematicians is to discover and to communicate certain truths. *Mathematics* is the language of mathematicians, and a *proof* is a method of communicating a mathematical truth to another person who also "speaks" the language. A remarkable property of the language of mathematics is its precision. Properly presented, a proof contains no ambiguity—there will be no doubt about its correctness. Unfortunately, many proofs that appear in textbooks and journal articles are presented for someone who already knows the language of mathematics. Thus, to understand and present a proof, you must learn a new language, a new method of thought. This book explains much of the basic grammar, but as in learning any new language, a lot of practice is needed to become fluent.

1.1 THE OBJECTIVES OF THIS BOOK

The approach taken here is to categorize and to explain the various **proof techniques** that are used in *all* proofs, regardless of the subject matter. One objective is to teach you how to read and understand a written proof by identifying the techniques that are used. Learning to do so enables you to study almost any mathematical subject on your own, a desirable goal in itself.

A second objective is to teach you to develop and to communicate your own proofs of known mathematical truths. Doing so requires you to use a certain amount of creativity, intuition, and experience. Just as there are many ways to express the same idea in any language, so there are different proofs for the same mathematical fact. The techniques presented here are designed to get you started and to guide you through a proof. Consequently, this book describes not only *how* the techniques work but also *when* each one is likely to be used and *why*. Often you will be able to choose a correct technique based on the form of the problem under consideration. Therefore, when attempting to create your own proof, *learn to select a technique consciously* before wasting hours trying to figure out what to do. The more aware you are of your thought processes, the better.

The ultimate objective, however, is to use your newly acquired skills and language to discover and communicate previously unknown mathematical truths. The first step in this direction is to reach the level of being able to read proofs and develop your own proofs of already-known facts. This alone will give you a much deeper and richer understanding of the mathematical universe around you.

Anyone with a good knowledge of high school mathematics can read this book. Advanced students who have seen proofs before can read the first two chapters, skip to the summary Chapter 13 and subsequently read the two appendices to see how all the techniques fit together. Each chapter on a particular technique also contains a brief summary at the end that describes how and when to use the technique. The remainder of this chapter explains the types of relationships to which proofs are applied. Additional books on proofs and advanced mathematical reasoning are listed in the bibliography at the end of this book.

1.2 WHAT IS A PROOF?

In mathematics, a **statement** is a sentence that is either true or false. Some examples follow:

1. Two different lines in a plane are either parallel or intersect at exactly one point.

2. $1 = 0$.

3. $3x = 5$ and $y = 1$.

4. $x \not> 0$ (x is not greater than 0).

5. There is an angle t such that $\cos(t) = t$.

Observe that statement (1) is always true, (2) is always false, and statements (3) and (4) are either true or false, depending on the value of a variable. For this reason, (3) and (4) are called **conditional statements**.

It is perhaps not as obvious that statement (5) is always true. It therefore becomes necessary to have some method for *proving* that such statements

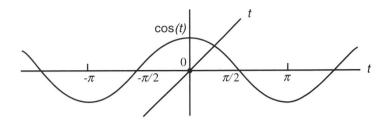

Fig. 1.1   A proof that there is an angle t such that $\cos(t) = t$.

are true. In other words, a **proof** is a convincing argument expressed in the language of mathematics. In this and other books, proofs are often given for what seem to be obviously true statements. One reason for doing so is to provide examples that are easy to follow so that you can develop techniques for proving more difficult statements. Another reason is that some apparently "obvious" statements are, in fact, false. You will know that a statement is true only when you have *proved* it to be true.

A proof should contain enough mathematical details to be convincing to the person(s) to whom the proof is addressed. A proof of statement (5) that is meant to convince a mathematics professor might consist of nothing more than Figure 1.1; whereas a proof directed toward a high school student would require more details, perhaps even the definition of cosine. Your proofs should contain enough details to be convincing to someone else at your own mathematical level (for example, a classmate). It is the lack of sufficient detail that often makes a proof difficult to read and understand. One objective of this book is to teach you to decipher "condensed" proofs that are likely to appear in textbooks and other mathematical literature.

Given two statements A and B, each of which may be either true or false, a fundamental problem of interest in mathematics is to show that the following statement—called an **implication**—is true:

If A is true, then B is true.

One reason for wanting to prove that an implication is true is when B is a statement that you would like to be true but whose truth is not easy to verify. In contrast, suppose that A is a statement whose truth is relatively easy to verify. If you have proved that "If A is true, then B is true," and if you can verify that A is in fact true, then you will know that B is true.

A proof is a formal method for convincing yourself (and others) that "If A is true, then B is true." To do a proof, you must know exactly what it means to show this. Statement A is called the **hypothesis** and B the **conclusion**. For brevity, the statement "If A is true, then B is true" is shortened to "If A, then B" or simply "A **implies** B." Mathematicians have developed a symbolic shorthand notation. For instance, a mathematician would write "$A \Rightarrow B$" instead of "A implies B." For the most part, textbooks do not use

Table 1.1   The Truth of "*A* Implies *B*."

A	*B*	*A* implies *B*
True	True	True
True	False	False
False	True	True
False	False	True

the symbolic notation, but teachers often do, and eventually you might find it useful, too. Therefore, the appropriate symbols are included in this book but are not used in the proofs. A complete list of symbols is presented in the glossary at the end of this book.

The conditions under which "*A* implies *B*" are true depend on whether *A* and *B* themselves are true. Thus, there are four possible cases to consider:

1. *A* is true and *B* is true.

2. *A* is true and *B* is false.

3. *A* is false and *B* is true.

4. *A* is false and *B* is false.

Suppose, for example, that your friend made the statement, "If it rains, then Mary brings her umbrella." Here, statement *A* is "it rains" and *B* is "Mary brings her umbrella." To determine when the statement "*A* implies *B*" is false, ask yourself in which of the four foregoing cases you would be willing to call your friend a liar. In the first case (that is, when it does rain and Mary does bring her umbrella), your friend has told the truth. In the second case, it has rained, and yet Mary did not bring her umbrella, as your friend said she would. Here your friend has not told the truth. In cases 3 and 4, it does not rain. You would not really want to call your friend a liar in the case of no rain because your friend said that something would happen only if it did rain. Thus, the statement "*A* implies *B*" is true in each of the four cases except the second one, as summarized in Table 1.1.

Table 1.1 is an example of a **truth table**, which is a method for determining when a complex statement (in this case, "*A* implies *B*") is true by examining all possible truth values of the individual statements (in this case, *A* and *B*). Other examples of truth tables appear in Chapter 3.

According to Table 1.1, when trying to show that "*A* implies *B*" is true, you might attempt to determine the truth of *A* and *B* individually and then use the appropriate row of the table to determine the truth of "*A* implies *B*." For example, to determine the truth of the statement,

if $1 < 2$, then $4 < 3$,

you can easily see that the hypothesis A (that is, $1 < 2$) is true and the conclusion B (that is, $4 < 3$) is false. Thus, using the second row of Table 1.1 (corresponding to A being true and B being false) you can conclude that in this case the statement "A implies B" is false. Similarly, the statement,

$$\text{if } 2 < 1, \text{ then } 3 < 4,$$

is true according to the third row of the table because A (that is, $2 < 1$) is false and B (that is, $3 < 4$) is true.

Now suppose you want to prove that the following statement is true:

$$\text{if } x > 2, \text{ then } x^2 > 4.$$

The difficulty with using Table 1.1 for this example is that you cannot determine whether A (that is, $x > 2$) and B (that is, $x^2 > 4$) are true or false because the truth of the conditional statements A and B depend on the variable x, whose value is not known. Nonetheless, you can still use Table 1.1 by reasoning as follows:

> Although I do not know the truth of A, I do know that A must be either true or false. Let me assume, for the moment, that A is false (subsequently, I will consider what happens when A is true). When A is false, either the third or the fourth row of Table 1.1 is applicable and, in either case, the statement "A implies B" is true—thus I would be done. Therefore, I need only consider the case in which A is true.

> When A is true, either the first or the second row of Table 1.1 is applicable. However, because I want to prove that "A implies B" is true, I need to be sure that the first row of the truth table is applicable, and this I can do by establishing that B is true.

From the foregoing reasoning, when trying to prove that "A implies B" is true, *you can assume that A is true; your job is to conclude that B is true.*

Note that a proof of the statement "A implies B" is not an attempt to verify whether A and B themselves are true but rather to show that B is a logical result of having assumed that A is true. Your ability to show that B is true depends on the fact that you have assumed A to be true; ultimately, you have to discover the relationship between A and B. Doing so requires a certain amount of creativity. The techniques presented here are designed to get you started and guide you along the path.

Summary

Hereafter, A and B are statements that are either true or false. The problem is to prove that "A implies B" is true. To do so,

1. Assume that A is true.

2. Use this assumption to reach the conclusion that B is true.

Exercises

Note: Solutions to exercises marked with a B are in the back of this book. Solutions to exercises marked with a W are located on the World Wide Web at http://www.wiley.com/college/solow/.

^{B}**1.1**   Which of the following are mathematical statements?

   a. $ax^2 + bx + c = 0$.

   b. $(-b + \sqrt{b^2 - 4ac})/(2a)$.

   c. Triangle XYZ is similar to triangle RST.

   d. $3 + n + n^2$.

   e. $\sin(\pi/2) < \sin(\pi/4)$.

   f. For every angle t, $\sin^2(t) + \cos^2(t) = 1$.

1.2   Which of the following mathematical statements are true?

   a. The cube root of any integer is a real number.

   b. For every angle t, $\sec^2(t) - \tan^2(t) = 1$.

   c. $x^2 + y^2 > 1$ (where x and y are real numbers).

   d. If $x > 0$, then $\log_7(x) > 0$ (where x is a real number).

^{B}**1.3**   For each of the following problems, identify the hypothesis (that is, what you can assume is true) and the conclusion (that is, what you are trying to show is true).

   a. If the right triangle XYZ with sides of lengths x and y and hypotenuse of length z has an area of $z^2/4$, then the triangle XYZ is isosceles.

   b. n is an even integer $\Rightarrow n^2$ is an even integer.

   c. It is possible to solve the two linear equations $ax + by = e$ and $cx + dy = f$ for x and y when a, b, c, d, e, and f are real numbers with $ad - bc \neq 0$.

^{W}**1.4**   For each of the following problems, identify the hypothesis (that is, what you can assume is true) and the conclusion (that is, what you are trying to show is true).

a. The sum of the first n positive integers is $n(n+1)/2$.

b. r is irrational if r is real and satisfies $r^2 = 2$.

c. If p and q are positive real numbers with $\sqrt{pq} \neq (p+q)/2$, then $p \neq q$.

d. When x is a real number, the minimum value of $x(x-1)$ is at least $-1/4$.

1.5   "If I do not get my car fixed, I will miss my job interview," says Jack. Later, you come to know that Jack's car was repaired but that he missed his job interview. Was Jack's statement true or false? Explain.

^{B}**1.6**   Using Table 1.1, determine the conditions on the hypothesis A and conclusion B under which the following statements are true and false and give your reason.

a. If $2 > 7$, then $1 > 3$.

b. If $2 < 7$, then $1 < 3$.

c. If $x = 3$, then $1 < 2$.

d. If $x = 3$, then $1 > 2$.

1.7   Suppose someone says to you that the following statement is true: "If Jack is younger than his father, then Jack will not lose the contest." Using Table 1.1, did Jack win the contest? Why or why not? Explain.

^{B}**1.8**   If you are trying to prove that "A implies B" is true and you know that B is false, do you want to show that A is true or false? Explain.

1.9   By considering what happens when A is true and when A is false, it was decided that when trying to prove the statement "A implies B" is true, you can assume that A is true and your goal is to show that B is true. Use the same type of reasoning to derive another approach for proving that "A implies B" is true by considering what happens when B is true and when B is false.

^{B}**1.10**   Using Table 1.1, prepare a truth table for the statement "A implies (B implies C)."

^{W}**1.11**   Using Table 1.1, prepare a truth table for the statement "(A implies B) implies C."

1.12   Suppose you want to show that "A implies B" is false.

a. According to Table 1.1, how should you do this? What should you try to show about the truth of A and B?

b. Apply your answer to part (a) to show that the following statement is false: "If $x > 0$, then $\log_7(x) > 0$."

$$\mathcal{2}$$

The Forward-Backward Method

The purpose of this chapter is to describe one of the fundamental proof techniques: the **forward-backward method**. Special emphasis is given to the material of this chapter because all other proof techniques rely on this method.

The first step in proving that "If A, then B" is true requires recognizing the statements A and B. In general, everything after the word "if" and before the word "then" is A. Everything after the word "then" is B. Alternatively, everything that you are assuming to be true (that is, the hypothesis) is A; everything that you are trying to prove (that is, the conclusion) is B. Consider the following example.

Proposition 1 *If the right triangle XYZ with sides of lengths x and y and hypotenuse of length z has an area of $z^2/4$, then the triangle XYZ is isosceles (see Figure 2.1).*

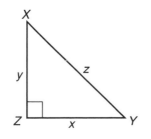

Fig. 2.1 The right triangle XYZ .

Analysis of Proof.   In this example you have the statements:

> **A:** The right triangle XYZ with sides of lengths x and y and hypotenuse of length z has an area of $z^2/4$.

> **B:** The triangle XYZ is isosceles.

Recall from the discussion at the end of Chapter 1 that, when proving "A implies B," you can assume that A is true and you must somehow use this information to reach the conclusion that B is true. In attempting to reach the conclusion that B is true, you will go through a **backward process**. When you make specific use of the information contained in A, you will go through a **forward process**. Both of these processes are described in detail now.

2.1   THE BACKWARD PROCESS

In the backward process you begin by asking, "How or when can I conclude that the statement B is true?" The very manner in which you phrase this question is critical because you must eventually provide an answer. You should pose the question in an abstract way. For Proposition 1, a correct abstract question is: "How can I show that *a* triangle is isosceles?" While it is true that you want to show that the specific triangle XYZ is isosceles, by asking the *abstract* question, you call on your general knowledge of triangles, clearing away irrelevant details (such as the fact that the triangle is called XYZ instead of ABC), thus allowing you to focus on those aspects of the problem that really seem to matter. The question obtained from statement B in such problems is referred to here as the **key question**. A properly posed key question should contain no symbols or other notation (except for numbers) from the specific problem under consideration. The key to many proofs is formulating a correct key question.

Once you have posed the key question, the next step in the backward process is to provide an answer. Returning to Proposition 1, how can you show that a triangle is isosceles? Certainly one way is to show that two of its sides have equal length. Referring to Figure 2.1, you should show that $x = y$. Observe that answering the key question is a two-phase process:

How to Answer a Key Question

1. First, give an abstract answer that contains no symbols from the specific problem. (For example, to show that *a* triangle is isosceles, show that two of *its* sides have equal length.)

2. Next, apply this abstract answer to the specific problem using appropriate notation. [For example, to show that two of *its* sides have equal length means to show that $x = y$ (not that $x = z$ or $y = z$)].

The process of asking the key question, providing an abstract answer, and then applying that answer to the specific problem constitutes one step of the backward process.

The backward process has given you a new statement, $B1$, with the property that if you could show that $B1$ is true, then B would be true. For the current example, the new statement is

B1: $x = y$.

If you can show that $x = y$, then the triangle XYZ is isosceles.

Once you have the statement $B1$, all of your efforts must now be directed toward reaching the conclusion that $B1$ is true, for then it will follow that B is true. How can you show that $B1$ is true? Eventually you will have to make use of the assumption that A is true. When solving this problem, you would most likely do so now but, for the moment, let us continue working backward from the new statement $B1$. This will illustrate some of the difficulties that can arise in the backward process. Can you pose a key question for $B1$?

Because x and y are the lengths of two sides of a triangle, a reasonable key question is, "How can I show that the lengths of two sides of a triangle are equal?" A second perfectly reasonable key question is, "How can I show that two real numbers are equal?" After all, x and y are also real numbers. One of the difficulties that can arise in the backward process is the possibility of more than one key question. Choosing the correct one is more of an art than a science. In fortunate circumstances, there will be only one obvious key question; in other cases you may have to proceed by trial and error. This is where your intuition, insight, creativity, experience, diagrams, and graphs can play an important role. One general guideline is to let the information in A (which you are assuming to be true) help you to choose the key question, as is done in this case.

Regardless of which question you finally settle on, the next step is to provide an answer, first in the abstract and then for the specific problem. Can you do this for the two foregoing key questions posed for $B1$? For the first one, you might show that two sides of a triangle have equal length by showing that the angles opposite them are equal. For the triangle XYZ in Figure 2.1, this would mean you have to show that angle X equals angle Y. A cursory examination of the contents of statement A does not seem to provide much information concerning those angles of triangle XYZ. For this reason, the other key question is chosen.

Now you are faced with the question, "How can I show that two real numbers (namely, x and y) are equal?" One answer to this question is to show that the difference of the two numbers is 0. Applying this answer to the specific statement $B1$ means you would have to show that $x - y = 0$. Unfortunately, there is another perfectly acceptable answer: show that the first number is less than or equal to the second number and also that the second number is less than or equal to the first number. Applying this answer to the specific statement $B1$, you would have to show that $x \le y$ and $y \le x$. Thus, a second difficulty can arise in the backward process: even if you choose the correct key question, there may be more than one answer. Moreover, you might choose an answer that will not permit you to complete the proof. For instance, associated with the key question, "How can I show that a triangle is isosceles?" is the answer, "Show that the triangle is equilateral." Of course it is impossible to show that triangle XYZ in Proposition 1 is equilateral because one of its angles is 90 degrees.

Returning to the key question associated with $B1$, "How can I show that two real numbers (namely, x and y) are equal?" suppose, for the sake of argument, that you choose the answer of showing that their difference is 0. Once again, the backward process has given you a new statement, $B2$, with the property that if you could show that $B2$ is true, then in fact $B1$ would be true, and hence so would B. Specifically, the new statement is:

B2: $x - y = 0$.

Now all of your efforts must be directed toward reaching the conclusion that $B2$ is true. You must ultimately make use of the information in A but, for the moment, let us continue the backward process applied to $B2$.

One key question associated with $B2$ is, "How can I show that the difference of two real numbers is 0?" After some reflection, it may seem that there is no reasonable answer to this question. Yet another problem can arise in the backward process: the key question might have no apparent answer. Do not despair—all is not lost. Remember that when proving "A implies B," you are allowed to assume that A is true. It is now time to make use of this fact.

2.2   THE FORWARD PROCESS

The forward process involves deriving from the statement A, which you assume is true, some other statement, $A1$, that you know is true as a result of A being true. It should be emphasized that the statements derived from A are not haphazard. Rather, they are directed toward linking up with the last statement obtained in the backward process. Let us return to Proposition 1, keeping in mind that the last statement in the backward process is,

B2: $x - y = 0$.

For Proposition 1, the statement A is, "The right triangle XYZ with sides of length x and y and hypotenuse of length z has an area of $z^2/4$." One fact that you know (or should know) as a result of A being true is that $xy/2 = z^2/4$, because the area of a triangle is one-half the base times the height—in this case, $xy/2$. So you have obtained the new statement,

A1: $xy/2 = z^2/4$.

Another useful statement follows from A by the Pythagorean theorem because XYZ is a right triangle, so you also have:

A2: $x^2 + y^2 = z^2$.

With the forward process, you can also combine and use the new statements to produce more true statements. For instance, it is possible to combine $A1$ and $A2$ by replacing z^2 in $A1$ with $x^2 + y^2$ from $A2$, obtaining the statement,

A3: $xy/2 = (x^2 + y^2)/4$.

One of the problems with the forward process is that it is also possible to generate useless statements, for instance, "angle X is less than 90 degrees." While there are no specific guidelines for producing new statements, keep in mind that the forward process is directed toward obtaining the statement $B2 : x - y = 0$, which was the last one derived in the backward process. The fact that $B2$ does not contain the quantity z^2 is the reason that z^2 was eliminated from $A1$ and $A2$ to produce $A3$.

Continuing with the forward process, you should attempt to make $A3$ look more like $B2$ by rewriting. For instance, you can multiply both sides of $A3$ by 4 and subtract $2xy$ from both sides to obtain,

A4: $x^2 - 2xy + y^2 - 0$.

Factoring $A4$ yields

A5: $(x - y)^2 = 0$.

One of the most common steps in the forward process is to rewrite statements in different forms, as is done in obtaining $A4$ and $A5$. For Proposition 1, the final step in the forward process (and in the entire proof) is to realize from $A5$ that if the square of a number (namely, $x - y$) is 0, then the number itself is 0, thus obtaining precisely the statement $B2 : x - y = 0$. The proof is now complete because you have successfully used the assumption that A is true to reach the conclusion that $B2$, and hence B, is true. The steps and reasons are summarized in Table 2.1.

It is interesting to note that the forward process ultimately produced the elusive answer to the key question associated with $B2$, "How can I show that the difference of two real numbers is 0?" In this case, the answer is to show that the square of the difference is 0 (see $A5$ in Table 2.1).

2.3   READING PROOFS

In general it is not practical to write the entire thought process that goes into a proof, for this requires too much time, effort, and space. Rather, a highly condensed version is usually presented and often makes little or no reference

Table 2.1   Proof of Proposition 1.

Statement	Reason
A: Area of XYZ is $z^2/4$	Given
$A1 : xy/2 = z^2/4$	Area = (base)(height)/2
$A2 : x^2 + y^2 = z^2$	Pythagorean theorem
$A3 : xy/2 = (x^2 + y^2)/4$	Substitute $A2$ into $A1$
$A4 : x^2 - 2xy + y^2 = 0$	From $A3$ by algebra
$A5 : (x - y)^2 = 0$	Factor $A4$
$B2 : x - y = 0$	From $A5$ by algebra
$B1 : x = y$	Add y to both sides of $B2$
B: XYZ is isosceles	Because $B1$ is true

to the backward process. For Proposition 1, the condensed proof might go something like this.

Proof of Proposition 1.   From the hypothesis and the formula for the area of a right triangle, the area of $XYZ = xy/2 = z^2/4$. By the Pythagorean theorem, $x^2 + y^2 = z^2$ and, on substituting $x^2 + y^2$ for z^2 and performing some algebraic manipulations, one obtains $(x - y)^2 = 0$. Hence $x = y$, and so the triangle XYZ is isosceles. □

In the foregoing proof, the sentence, "From the hypothesis ..." indicates that the author is working forward. Also, the □ or some equivalent symbol is usually used to indicate the end of the proof. Sometimes the letters Q.E.D. are used as they stand for the Latin words *quod erat demonstrandum*, meaning "which was to be demonstrated."

Sometimes the condensed proof is partly backward and partly forward. For example:

Proof of Proposition 1.   The statement is proved by establishing that $x = y$, which in turn is done by showing that $(x-y)^2 = x^2 - 2xy + y^2 = 0$. But the area of the triangle is $xy/2 = z^2/4$ so that $2xy = z^2$. By the Pythagorean theorem, $z^2 = x^2 + y^2$ and hence $x^2 + y^2 = 2xy$, or, $x^2 - 2xy + y^2 = 0$. □

The proof can also be written entirely from the backward process. Although slightly unnatural, this version is worth seeing.

Proof of Proposition 1.   To reach the conclusion, it is shown that $x = y$ by verifying that $(x - y)^2 = x^2 - 2xy + y^2 = 0$, or equivalently, that $x^2 + y^2 = 2xy$. This is established by showing that $2xy = z^2$, for the Pythagorean theorem states that $x^2 + y^2 = z^2$. To see that $2xy = z^2$, or equivalently, that $xy/2 = z^2/4$, note that $xy/2$ is the area of the triangle and is equal to $z^2/4$ by hypothesis, thus completing the proof. □

Proofs found in research articles are very condensed, giving little more than a hint of how to do the proof. For example:

Proof of Proposition 1.   The hypothesis together with the Pythagorean theorem yield $x^2 + y^2 = 2xy$ and hence $(x - y)^2 = 0$. Thus the triangle is isosceles, as required. □

Note that the word "hence" effectively conceals the reason that $(x - y)^2 = 0$. Is that reason algebraic manipulation (as we know it is) or something else?

Reasons Why Reading a Condensed Proof Is Challenging

From these examples, you can see that there are several reasons why reading a condensed proof is challenging:

1. The steps of the proof are not always presented in the same order in which they were performed when the proof was done (see, for example, the first condensed proof of Proposition 1 given above).

2. The names of the techniques are often omitted (for instance, in the foregoing condensed proofs, no mention is made of the forward and backward processes or the key question).

3. Several steps of the proof are often combined into a single statement with little or no explanation (as is done in the last condensed proof of Proposition 1 given above).

Steps for Reading a Condensed Proof

You should strive toward the ability to read and to dissect a condensed proof. To do so, you have to discover the thought processes that went into the proof, which you can do by following these steps.

1. Determine which techniques are used (because the forward-backward method is not the only one available).

2. Verify all of the steps involved by filling in missing details.

The more condensed the proof, the harder this process is. When an author writes, "It is easy to see that ...," or "Clearly, ..., " you can assume it will take you quite some time to fill in the missing details. In this book, reading proofs is made easier because preceding each condensed proof is an analysis that describes the techniques, methodology, and reasoning involved in doing the proof. Additional suggestions and practice on reading condensed proofs are given throughout the book and in various exercises.

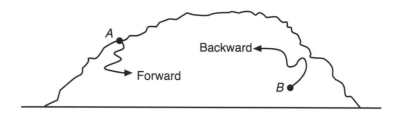

Fig. 2.2   Finding a needle in a haystack.

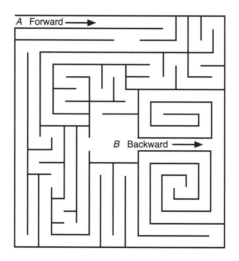

Fig. 2.3   The maze.

Summary

To use the forward-backward method for proving that "A implies B," begin with the statement B that you are trying to conclude is true. Through the backward process of asking and answering the key question, derive a new statement, $B1$, with the property that if $B1$ is true, then so is B. All efforts are now directed toward establishing that $B1$ is true. To that end, apply the backward process to $B1$, obtaining a new statement, $B2$, with the property that if $B2$ is true, then so is $B1$ (and hence B). The backward process is motivated by the fact that A is assumed to be true. Continue working backward until either you obtain the statement A (in which case the proof is finished) or until you can no longer pose and/or answer the key question fruitfully. In the latter case, it is time to start the forward process, in which you derive a sequence of statements from A that are necessarily true as a result of assuming that A is true. Remember that the goal of the forward process is to obtain precisely the last statement you have in the backward process, at which time the proof is complete.

These two processes are easily remembered by thinking of the statement B as a needle in a haystack. When you work forward from the assumption that A is true, you start somewhere on the outside of the haystack and try to find the needle. In the backward process, you start at the needle and try to work your way out of the haystack toward the statement A (see Figure 2.2).

Another way of remembering the forward-backward method is to think of a maze in which A is the starting point and B is the ending point (see Figure 2.3). You may have to alternate several times between the forward and backward processes because there are likely to be several false starts and blind alleys.

As a general rule, the forward-backward method is probably the first technique to try on a problem unless you have reason to use a different approach based on the form of B, as described shortly. In any case, you will gain much insight into the relationship between A and B.

To read a condensed proof, you have to discover the thought processes that went into the proof. To do so, (a) determine which techniques are used and (b) verify all of the steps involved by filling in missing details.

Exercises

Note: Solutions to exercises marked with a B are in the back of this book. Solutions to exercises marked with a W are located on the World Wide Web at http://www.wiley.com/college/solow/.

Note: All proofs should contain an analysis of proof and a condensed version. Definitions for all mathematical terms are provided in the glossary at the end of the book.

^{W}**2.1**   Explain the difference between the forward and backward processes. Describe how each one works and what can go wrong. How are the two processes related to each other?

2.2   When formulating the key question, should you look at the last statement in the forward or backward process? When answering the key question, should you be guided by the last statement in the forward or backward process?

^{B}**2.3**   Consider the problem of proving that, "If x is a real number, then the maximum value of $-x^2 + 2x + 1$ is greater than or equal to 2." Which of the following key questions is incorrect? Explain.

    a. How can I show that the maximum value of a parabola is greater than or equal to a number?

    b. How can I show that a number is less than or equal to the maximum value of a polynomial?

    c. How can I show that the maximum value of the function $-x^2 + 2x + 1$ is greater than or equal to a number?

d. How can I show that a number is less than or equal to the maximum of a quadratic function?

2.4   For the key question, "How can I show that two lines in a plane are parallel?" which of the following answers is incorrect? Explain.

a. Show that the slopes of the two lines are the same.

b. Show that the each of the two lines is parallel to a third line.

c. Show that each of the two lines is perpendicular to a third line.

d. Show that the lines are on opposite sides of a quadrilateral.

^{B}**2.5**   Consider the problem of showing that, "If

$$R = \{\text{real numbers } x : x^2 - x \leq 0\},$$
$$S = \{\text{real numbers } x : -(x - 1)(x - 3) \leq 0\}, \quad \text{and}$$
$$T = \{\text{real numbers } x : x \geq 1\},$$

then R intersect S is a subset of T." Which of the following key questions is correct, and why? Explain what is wrong with the other choices.

a. How can I show that a set is a subset of another set?

b. How can I show that R intersect S is a subset of T?

c. How can I show that every point in R intersect S is greater than or equal to 1?

d. How can I show that the intersection of two sets has a point in common with another set?

2.6   Suppose you are trying to prove that, "If l_1 and l_2 are parallel lines intersected by a third line l_3, then the smallest angle, S, formed by the intersection of l_1 and l_3 is equal to the smallest angle, T, formed by the intersection of l_2 and l_3." What is wrong with the key question, "How can I show that two parallel lines intersect a third line?"

^{B}**2.7**   For each of the following problems, list as many key questions as you can (at least two). Be sure your questions contain no symbols or notation from the specific problem.

a. If l_1 and l_2 are tangent lines to a circle C at the two endpoints e_1 and e_2 of a diameter d, respectively, then l_1 and l_2 are parallel.

b. If f and g are continuous functions, then the function $f + g$ is continuous. (Note: Continuity is a property of a function.)

^{W}**2.8**   For each of the following problems, list as many key questions as you can (at least two). Be sure your questions contain no symbols or notation from the specific problem.

  a. If n is an even integer, then n^2 is an even integer.

  b. If n is a given integer satisfying $-3n^2 + 2n + 8 = 0$, then $2n^2 - 3n = -2$.

2.9   For each of the following problems, list as many key questions as you can (at least two). Be sure your questions contain no symbols or notation from the specific problem.

  a. If a and b are nonnegative real numbers, then $a^2 + b^2 \le (a + b)^2$.

  b. If $y = m_1 x + b_1$ and $y = m_2 x + b_2$ are the equations of two lines for which $m_1 = m_2$, then the two lines are parallel.

  c. If RST is an isosceles triangle with sides RS, ST, and TR, and SU is a perpendicular bisector of RT, then $\overline{RS} = \overline{ST}$.

  d. If R and T are the sets described in Exercise 2.5, then R intersect T is a singleton (a set with only one element).

^{B}**2.10**   For each of the following key questions, list as many answers as you can (at least three).

  a. How can I show that two real numbers are equal?

  b. How can I show that two triangles are congruent?

^{W}**2.11**   For each of the following key questions, list as many answers as you can (at least three).

  a. How can I show that two lines are perpendicular?

  b. How can I show that two sets are equal?

2.12   For each of the following key questions, list as many answers as you can (at least three).

  a. How can I show that two lines are parallel?

  b. How can I show that a set is a subset of another set?

^{W}**2.13**   For each of the following problems, (1) pose a key question, (2) answer the question abstractly, and (3) apply your answer to the specific problem.

a. If a, b, and c are real numbers for which $a > 0$, $b < 0$, and $b^2 - 4ac = 0$, then the solution to the equation $ax^2 + bx + c = 0$ is positive.

b. In the following diagram, if SU is a perpendicular bisector of RT, and $\overline{RS} = 2\overline{RU}$, then triangle RST is equilateral.

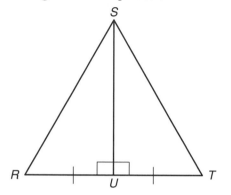

2.14   For the triangles RSU and SUT in the figure for Exercise 2.13(b), suppose you have asked the key question, "How can I show that two triangles are congruent?" What is wrong with the answer, "Show that $\angle RUS = \angle SUT$, $\angle RSU = \angle UST$, and $\angle SRU = \angle STU$"?

^{B}**2.15**   For each of the following hypotheses, list as many statements as you can (at least two) that are a result of applying the forward process one step from the hypothesis.

a. The real number x satisfies $x^2 - 3x + 2 < 0$.

b. The sine of angle X in triangle XYZ of Figure 2.1 is $1/\sqrt{2}$.

c. The circle C consists of all values for x and y that satisfy the equation $(x - 3)^2 + (y - 2)^2 = 25$.

2.16   For each of the following hypotheses, list as many statements as you can (at least two) that are a result of applying the forward process one step from the hypothesis.

a. The rectangle $ABCD$ is a square.

b. The integer $n^2 - 1$ is odd.

c. The line $y = 3x - 1$ is tangent to the function $x^2 + x$.

^{B}**2.17**   Consider the problem of proving that, "If x and y are real numbers such that $x^2 + 6y^2 = 25$ and $y^2 + x = 3$, then $y = 2$." In working forward from the hypothesis, which of the following is not valid? Explain.

a. $y^2 = 3 - x$.

b. $y^2 = 25/6 - (x/\sqrt{6})^2$.

c. $(3 - y^2)^2 + 6y^2 - 25 = 0$.

d. $x + 5 = -6y^2/(x - 5)$.

2.18   Suppose you are trying to prove that, "If R is a subset of S and S is a subset of T, then R is a subset of T." What is wrong with the following statement in the forward process: "Because R is a subset of S, it follows that every element of S is in also an element of R."

^{B}**2.19**   Consider the problem of proving that, "If x and y are nonnegative real numbers that satisfy $x + y = 0$, then $x = 0$ and $y = 0$."

   a. For the following condensed proof, write an analysis indicating the forward and backward steps and the key questions and answers.

> **Proof.** First, it will be shown that $x \leq 0$, for then, because $x \geq 0$ by the hypothesis, it must be that $x = 0$. To see that $x \leq 0$, by the hypothesis, $x + y = 0$, so $x = -y$. Also, because $y \geq 0$, it follows that $-y \leq 0$ and hence $x = -y \leq 0$. Finally, to see that y$= 0$, because $x = 0$ and $x + y = 0$, it follows that $0 = x + y = 0 + y = y$. $\square$

   b. Rewrite the condensed proof of part (a) entirely from the backward process.

^{W}**2.20**   For the problem of proving that, "If n is an integer greater than 2, a and b are the lengths of the legs of a right triangle, and c is the length of the hypotenuse, then $c^n > a^n + b^n$," provide justification for each sentence in the following condensed proof.

> **Proof.** You have that $c^n = c^2 c^{n-2} = (a^2 + b^2)c^{n-2}$. Observing that $c^{n-2} > a^{n-2}$ and $c^{n-2} > b^{n-2}$, it follows that $c^n > a^2(a^{n-2}) + b^2(b^{n-2})$. Consequently, $c^n > a^n + b^n$. $\square$

2.21   Consider the problem of proving that, "If RST is the triangle in Exercise 2.13(b), then triangle SUR is congruent to triangle SUT." For the following condensed proof, write an analysis indicating the forward and backward steps, and the key questions and answers.

> **Proof.** It will be shown that $\overline{RU} = \overline{UT}$, for then you also have $\angle RUS = \angle TUS = 90^o$ and $\overline{SU} = \overline{SU}$. To that end, note that $\overline{RU} = \overline{UT}$ because SU is a perpendicular bisector of RT, and so the proof is complete. $\square$

^{W}**2.22**   Consider an alphabet consisting of the two letters s and t, together with the following rules for creating new "words" from old ones. (You can apply the rules in any order.)

(1) Double the current word (for example, sts could become $stssts$).

(2) Erase tt from the current word (for example, $stts$ could become ss).

(3) Replace sss in the current word by t (for example, $stsss$ could become stt).

(4) Add the letter t at the right end of the current word if the last letter is s (for example, tss could become $tsst$).

    a. Use the forward process to derive all possible words that can be obtained in three steps by repeatedly applying the above rules in any order to the initial word s.

    b. Apply the backward process one step to the word tst. Specifically, list all the words for which an application of one of the above rules would result in tst.

    c. Prove that "If s, then tst."

    d. Prove that "If s, then $ttst$."

^{B}**2.23**   Prove that if the right triangle XYZ in Figure 2.1 is isosceles, then the area of the triangle is $z^2/4$.

2.24   Prove that if XYZ is a triangle with $\angle X = 30^o$, $\angle Y = 60^o$, and $\angle Z = 90^o$ and W is the midpoint of the hypotenuse, then the line connecting W to Z divides the triangle XYZ into an equilateral triangle and an isosceles triangle.

^{W}**2.25**   Prove that the statement in Exercise 2.13(b) is true.

2.26   Prove that if the triangle RST in Exercise 2.13(b) is equilateral and SU is a perpendicular bisector of RT, then the area of triangle RUS is $\sqrt{3}\left(\overline{RS}\right)^2/4$.

3

On Definitions and Mathematical Terminology

In the previous chapter you learned the forward-backward method and saw the importance of formulating and answering the key question. One of the simplest yet most effective ways of answering a key question is through the use of a definition, as explained in this chapter. You will also learn some of the "vocabulary" of the language of mathematics.

3.1 DEFINITIONS

A **definition** in mathematics is an agreement, by all parties concerned, as to the meaning of a particular term. You have already come across a definition in Chapter 1. There, the statement "A implies B" was defined—and hence, agreed—to be true in all cases except when A is true and B is false. Nothing says that you must accept this definition as being correct. If you choose not to, then we will be unable to communicate regarding this particular idea.

Definitions are not made randomly. Usually they are motivated by a mathematical concept that occurs repeatedly. You can view a definition as an abbreviation that is agreed on for a particular concept. Take, for example, the notion of "a positive integer greater than one that is not divisible by any positive integer other than one and itself." This type of number is abbreviated (or defined) as a "prime." Surely it is easier to say "prime" than "a positive integer greater than one ...," especially if the concept comes up frequently. Several other examples of definitions follow:

Definition 1 *An integer n **divides** an integer m (written $n|m$) if $m = kn$, for some integer k.*

Definition 2 *A positive integer $p > 1$ is **prime** if the only positive integers that divide p are 1 and p.*

Definition 3 *A triangle is **isosceles** if two of its sides have equal length.*

Definition 4 *Two pairs of real numbers (x_1, y_1) and (x_2, y_2) are **equal** if $x_1 = x_2$ and $y_1 = y_2$.*

Definition 5 *An integer n is **even** if and only if the remainder on dividing n by 2 is 0.*

Definition 6 *An integer n is **odd** if and only if $n = 2k + 1$ for some integer k.*

Definition 7 *A real number r is **rational** if and only if r can be expressed as the ratio of two integers p and q in which the denominator q is not 0.*

Definition 8 *Two statements A and B are **equivalent** if and only if "A implies B" and "B implies A."*

Definition 9 *The statement A AND B (written $A \bigwedge B$) is true if and only if A is true and B is true.*

Definition 10 *The statement A OR B (written $A \bigvee B$) is true in all cases except when A is false and B is false.*

Observe that the words "if and only if" are used in some of the definitions, but often, "if" is used instead of "if and only if." Some terms, such as "set" and "point," are left undefined. One could possibly try to define a set as a collection of objects, but to do so is impractical because the concept of an "object" is too vague; one would then be led to ask for the definition of an "object," and so on, and so on. Such philosophical issues are beyond the scope of this book.

Using Definitions in Proofs

In the proof of Proposition 1, a definition is used to answer a key question. Recall the first one, which is, "How can I show that a triangle is isosceles?" Using Definition 3, to show that a triangle is isosceles, one shows that two of its sides have equal length. Definitions are equally useful in the forward process. For instance, if you know that an integer n is odd, then by Definition 6 you would know that $n = 2k + 1$, for some integer k. Using definitions to work forward and backward is a common occurrence in proofs.

It is often the case that there seem to be two possible definitions for the same concept. Take, for example, the notion of an even integer introduced in Definition 5. A second plausible definition for an even integer is "an integer that can be expressed as 2 times some other integer." There can be only one definition for a particular concept, so, when more possibilities exist, how do you select the definition and what happens to the other alternatives?

Because a definition is simply something agreed on, any one of the alternatives can be agreed on as the definition. Once the definition is chosen, it is advisable to establish the "equivalence" of the definition and the alternatives. For the case of an even integer, this is accomplished by using Definition 5 and the alternative to create the statements:

A: n is an integer whose remainder on dividing by 2 is 0.

B: n is an integer that can be expressed as 2 times some integer.

To establish that the definition is equivalent to the alternative, you must show that "A implies B" and "B implies A" (see Definition 8). Then you would know that if A is true (that is, n is an integer whose remainder on dividing by 2 is 0), then B is true (that is, n is an integer that can be expressed as 2 times some other integer). Moreover, if B is true then A is true, too.

The statement that A is equivalent to B is often written "A is true if and only if B is true," or, "A if and only if B." In mathematical notation, one would write "A iff B" or "$A \Leftrightarrow B$." Whenever you need to show that "A if and only if B," you must show that "A implies B" and "B implies A."

It is useful to establish that a definition is equivalent to an alternative. To see why, suppose that in some proof you derive the key question, "How can I show that an integer is even?" As a result of having obtained the equivalence of the two concepts, you now have two possible answers available. One is obtained from the definition: show that the remainder on dividing the integer by 2 is 0; the second answer comes from the alternative: show that the integer can be expressed as 2 times some other integer. Similarly, in the forward process, if you know that n is an even integer, then you would have two possible statements that are true as a result of this—the original definition and the alternative. While the ability to answer a key question (or to work forward) in more than one way can be a hindrance, as was the case in the proof of Proposition 1, it can also be advantageous, as is shown now.

Proposition 2 *If n is an even integer, then n^2 is an even integer.*

Analysis of Proof. Proceeding by the forward-backward method, you are led to the key question, "How can I show that an integer (namely, n^2) is even?" Choosing the alternative over the definition, you can answer this question by showing that

B1: n^2 can be expressed as 2 times some other integer.

The only question is, which integer. The answer comes from the forward process.

Because n is an even integer, using the alternative, n can be expressed as 2 times some other integer, say, k, that is,

A1: $n = 2k$.

Squaring both sides of $A1$ and rewriting by algebra yields

A2: $n^2 = (n)(n) = (2k)(2k) = 4k^2 = 2(2k^2)$.

Thus, it has been shown that n^2 is 2 times some other integer, that integer being $2k^2$, and this completes the proof. You could also prove this proposition by using Definition 5, but it is harder that way.

Proof of Proposition 2. Because n is an even integer, there is an integer k for which $n = 2k$. Consequently $n^2 = (2k)^2 = 2(2k^2)$, and so n^2 is an even integer. $\square$

A definition is one common method for working forward and for answering key questions. The more statements you can show are equivalent to the definition, the more ammunition you have available for the forward and backward processes; however, a large number of equivalent statements can also make it challenging to know exactly which one to use.

Notational Issues

Notational difficulties sometimes arise when using definitions in the forward and backward processes. This occurs when the definition uses one set of symbols and notation while the specific problem under consideration uses a second set of symbols and notation. When these two sets of symbols are completely distinct from each other, generally no confusion arises; however, care is needed when these two sets involve **overlapping notation**—that is, when the same symbol is used in both sets.

To illustrate, recall the foregoing Definition 1:

> Definition 1. An integer n **divides** an integer m (written $n|m$) if $m = kn$, for some integer k.

Suppose the conclusion of the problem you are working with is:

B: The integer p divides the integer q.

The key question associated with B is, "How can I show that one integer (namely, p) divides another integer (namely, q)?" Because the symbols p and q in B are distinct from those in the definition, you should have no trouble using the definition to obtain the following answer to the key question:

B1: $q = kp$, for some integer k.

Observe that $B1$ is obtained by **matching up the notation** in Definition 1 with that in B. That is, the symbol p in B is "matched" to the symbol n in the definition; the symbol q in B is "matched" to the symbol m in the definition. $B1$ is then obtained by replacing n everywhere in the definition with p and m everywhere with q.

This process of matching up the notation when using a definition is similar to a process you have seen when working with functions. To illustrate, suppose f is a function of one variable defined by $f(x) = x(x+1)$. To write $f(a+b)$, you "match" x to $(a+b)$, that is, you replace x everywhere by $(a+b)$ to obtain $f(a+b) = (a+b)(a+b+1)$.

To see how notational difficulties arise when using definitions, suppose the last statement in the backward process of the problem you are working with is:

B2: The integer k divides the integer n.

Once again, the key question associated with $B2$ is, "How can I show that one integer (namely, k) divides another integer (namely, n)?" A difficulty in applying Definition 1 to answer this question arises because of overlapping notation—the symbols k and n appear in both $B2$ and in the definition, but the symbols are used differently in each case. Observe how this overlapping notation makes it challenging to match up the notation in $B2$ with that in the definition.

When overlapping notation occurs, you can avoid notational errors by first rewriting the definition using a new set of symbols that do not overlap with the specific problem under consideration. Then when you apply the definition to the specific problem, the matching up of notation will be clear. For the foregoing example, you could rewrite the definition as follows so that it contains no overlapping notation with $B2$:

Definition 1. An integer a **divides** an integer b (written $a|b$) if $b = ca$, for some integer c.

Now when you apply this definition to answer the key question associated with $B2$, "How can I show that one integer (namely, k) divides another integer (namely, n)?" you should have no trouble matching up the notation correctly (match k to a and n to b) to obtain the following answer:

B3: $n = ck$, for some integer c.

With practice, you will find that there really is no need to rewrite a definition. This is because you will be able to match up notation correctly even when there is overlapping notation. Until you reach that point, however, it is advisable to rewrite the definition. It is critical to learn how to apply a definition correctly in the forward and backward processes because, if you make a mistake, the rest of the proof is incorrect.

3.2 USING PREVIOUS KNOWLEDGE

Just as you can use a definition in the forward and backward processes, so you can use a previously proven implication, as is shown now.

Proposition 3 *If the right triangle RST with sides of lengths r and s and hypotenuse of length t satisfies $t = \sqrt{2rs}$, then the triangle RST is isosceles (see Figure 3.1).*

Analysis of Proof.   The forward-backward method gives rise to the key question, "How can I show that a triangle (namely, RST) is isosceles?" One answer is to use Definition 3, but a second answer is provided by the conclusion of Proposition 1, which states that the triangle XYZ is isosceles. Perhaps the current triangle RST is also isosceles for the same reason as triangle XYZ. To find out, it is necessary to see if RST also satisfies the hypothesis of Proposition 1, as did triangle XYZ, for then RST will also satisfy the conclusion, and hence be isosceles.

In verifying the hypothesis of Proposition 1 for the triangle RST, it is first necessary to match up the current notation with that of Proposition 1 (just as is done when applying a definition). In this case, the corresponding lengths are $x = r$, $y = s$, and $z = t$. Thus, to check the hypothesis of Proposition 1 for the current problem, you must show that

B1: The area of triangle RST equals $t^2/4$,

or equivalently, because the area of triangle RST is $rs/2$, you must show that

B2: $rs/2 = t^2/4$.

The fact that $rs/2 = t^2/4$ is established by working forward from the current hypothesis that

A: $t = \sqrt{2rs}$.

To be specific, on squaring both sides of the equality in A and dividing by 4, you obtain the desired conclusion that

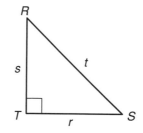

Fig. 3.1 The right triangle RST.

A1: $rs/2 = t^2/4$.

Do not forget to observe that the hypothesis of Proposition 1 also requires that the triangle RST be a right triangle, which of course it is, as stated in the current hypothesis.

Notice how much more challenging it would be to match up the notation if the current triangle had been labeled WXY with sides of length w and x and hypotenuse of length y. This overlapping notation can (and will) arise and, when it does, you should use the same technique as in the case of definitions, that is, you should rewrite the previous proposition with a set of symbols that do not overlap with the current problem.

In the condensed proof that follows, note the complete lack of reference to the matching of notation.

Proof of Proposition 3. By the hypothesis, $t = \sqrt{2rs}$, so $t^2 = 2rs$, or equivalently, $t^2/4 = rs/2$. Thus the area of the right triangle $RST = t^2/4$. As such, the hypothesis, and hence the conclusion, of Proposition 1 is true. Consequently, the triangle RST is isosceles. $\square$

3.3   MATHEMATICAL TERMINOLOGY

In dealing with proofs, there are four terms you will often come across in mathematics: *proposition, theorem, lemma,* and *corollary.* A **proposition** is a true statement of interest that you are trying to prove. Some propositions are subjectively considered to be extremely important, and these are referred to as **theorems**. When the proof of a theorem is long, the proof is often easier to communicate in "pieces." For example, when proving the statement "*A* implies *B*," it may first be convenient to show that "*A* implies *C*," then that "*C* implies *D*," and finally that "*D* implies *B*." Each of these supporting propositions would be presented separately and referred to as a *lemma*. In other words, a **lemma** is a preliminary proposition that is used in the proof of a theorem. Once a theorem is proved, it is often the case that certain propositions follow almost immediately as a result of knowing that the theorem is true. These are called **corollaries**.

Just as there are certain mathematical concepts that are accepted without a formal definition, so certain statements are accepted without a formal proof. These unproved statements are called **axioms**. One example of an axiom is, "the shortest distance between two points is a straight line." A further discussion of axioms is beyond the scope of this book.

Associated with a statement A is the statement $NOT\ A$ (sometimes written $\neg A$ or $\sim A$). The statement $NOT\ A$ is true when A is false, and vice versa. More is said about the NOT of a statement in Chapter 10.

Table 3.1   Statements Related to "*A* Implies *B*."

Statement	Alternate Written Form	Name of Statement
"*B* implies *A*"	$B \Rightarrow A$	converse
"*NOT A* implies *NOT B*"	$(\neg A) \Rightarrow (\neg B)$	inverse
"*NOT B* implies *NOT A*"	$(\neg B) \Rightarrow (\neg A)$	contrapositive

Given two statements A and B, you have already learned the meaning of the statement "*A* implies *B*." There are other ways of saying the statement "*A* implies *B*," for example:

1. Whenever A is true, B must also be true.

2. B follows from A.

3. B is a necessary consequence of A (meaning that if A is true, B is necessarily true also.)

4. A is sufficient for B (meaning that, if you want B to be true, it is enough to know that A is true.)

5. A only if B.

Three other statements closely related to "*A* implies *B*" are the **contrapositive statement**, the **converse statement**, and the **inverse statement**, as given in Table 3.1. You can use Table 1.1 to determine when each of these three statements is true. For instance, the contrapositive statement, "*NOT B* implies *NOT A*," is true in all cases except when the statement to the left of the word "implies" (namely, *NOT B*) is true and the statement to the right of the word "implies" (namely, *NOT A*) is false. In other words, the contrapositive statement is true in all cases except when B is false and A is true, as shown in Table 3.2.

Note, from Table 3.2, that the statement "*NOT B* implies *NOT A*" is true under the same conditions as "*A* implies *B*," that is, in all cases except when

Table 3.2   The Truth The Truth Table for "*NOT B* Implies *NOT A*."

A	B	$NOT\ B$	$NOT\ A$	$A \Rightarrow B$	$NOT\ B \Rightarrow NOT\ A$
True	True	False	False	True	True
True	False	True	False	False	False
False	True	False	True	True	True
False	False	True	True	True	True

A is true and B is false. This observation gives rise to a new proof technique known as the contrapositive method that is described in Chapter 9. In the exercises, you are asked to derive truth tables similar to Table 3.2 for the converse and inverse statements.

Summary

In this chapter, you have learned how definitions and previous propositions are used in the forward-backward method. A definition is used in the backward process to answer a key question and in the forward process to derive new statements. To use a definition, you must correctly match the notation of the current problem to that of the definition. If there is overlapping notation between the two, you might want to rewrite the definition using a different set of symbols.

To use a previous proposition to prove a new one, follow these steps:

1. Find a previous proposition whose conclusion is the same, except possibly for the notation, as the one under consideration.

2. Match up the current notation to that of the previous proposition. (If there is overlapping notation, you might first want to rewrite the previous one so that there is no overlapping notation.)

3. Verify that the hypothesis of the previous proposition is true for the current one. (The previous hypothesis written in terms of the notation of the current problem becomes the new statement to prove in the backward process.)

4. State that the conclusion of the previous proposition is true for the current problem. (Stating the previous conclusion in terms of the notation of the current problem becomes the next statement in the forward process.)

You have also learned many of the terms used in the language of mathematics: a proposition is a true statement that you are trying to prove; a theorem is an important proposition; a lemma is a preliminary proposition that is used in the proof of a theorem; and a corollary is a proposition that follows from a theorem. Now it is time to learn more proof techniques.

Exercises

Note: Solutions to exercises marked with a B are in the back of this book. Solutions to exercises marked with a W are located on the World Wide Web at http://www.wiley.com/college/solow/.

Note: All proofs should contain an analysis of proof and a condensed version. Definitions for all mathematical terms are provided in the glossary at the end of the book.

^{B}**3.1**   For each of the following conclusions, pose a key question. Then use a definition (1) to answer the question abstractly and (2) to apply the answer to the specific problem.

    a. If n is an odd integer, then n^2 is an odd integer.

    b. If s and t are rational numbers with $t \neq 0$, then s/t is rational.

    c. Suppose that a, b, c, d, e, and f are real numbers with the property that $ad - bc \neq 0$. If (x_1, y_1) and (x_2, y_2) are pairs of real numbers satisfying:

$$ax_1 + by_1 = e, \quad cx_1 + dy_1 = f,$$

$$ax_2 + by_2 = e, \quad cx_2 + dy_2 = f,$$

then (x_1, y_1) equals (x_2, y_2).

    d. If n is an integer greater than 1 for which $2^n - 1$ is prime, then n is prime.

    e. If $n - 1$, n, and $n + 1$ are three consecutive integers, then 9 divides the sum of their cubes.

3.2   For each of the following statements, obtain a new statement in the backward process by using a definition to answer the key question. When necessary, rewrite the definitions so that no overlapping notation occurs.

    a. m is prime.

    b. Triangle ABC is equilateral.

    c. p^2 is even.

    d. $\sqrt{n}$ is rational.

^{W}**3.3**   For each of the following hypotheses, use a definition to work forward one step.

    a. If n is an odd integer, then n^2 is an odd integer.

    b. If s and t are rational numbers with $t \neq 0$, then s/t is rational.

    c. If triangle RST is equilateral, then the area of RST is $\sqrt{3}/4$ times the square of the length of a side.

d. If the right triangle XYZ of Figure 2.1 on page 9 satisfies $\sin(X) = \cos(X)$, then triangle XYZ is isosceles.

e. If a, b, and c are integers for which $a|b$ and $b|c$, then $a|c$.

3.4   Use a definition to work forward from each of the following statements.

a. For sets R, S, and T, $R = S \cup T$.

b. For functions f and g, the function $f + g$ is convex, where $f + g$ is the function whose value at any point x is $f(x) + g(x)$.

c. For functions f and g and sets S and T, the function $f \geq g$ on $S \cap T$. (See the definition of greater-than-or-equal-to functions in the glossary.)

^{B}**3.5**   Write truth tables for the following statements.

a. The converse of "A implies B."

b. The inverse of "A implies B." How are (a) and (b) related?

^{W}**3.6**   Write truth tables for the following statements.

a. A OR B.

b. A AND B.

c. A AND NOT B.

d. $(NOT\ A)$ OR B. How is this related to the truth of "A implies B"?

^{B}**3.7**   Write the indicated statement for each of the following implications.

a. Write the contrapositive of the statement, "If n is an integer for which n^2 is even, then n is even."

b. Write the inverse of the statement, "If r is a real number such that $r^2 = 2$, then r is not rational."

c. Write the converse of the statement, "If the quadrilateral $ABCD$ is a parallelogram with one right angle, then $ABCD$ is a rectangle."

3.8   Suppose that A, B, and C are statements. Is the statement "A implies $(B$ OR $C)$" equivalent to the statement "$(A$ AND NOT $B)$ implies C"? Why or why not? Explain.

^{W}**3.9**   Prove that if n is an odd integer, then n^2 is an odd integer.

^{B}**3.10**   Use the proposition in Exercise 3.9 to prove that if a and b are consecutive integers, then $(a + b)^2$ is an odd integer.

3.11   Use Proposition 2 on page 25 to prove that if a and b are even integers, then $(a + b)^2$ is an even integer.

^{B}**3.12**   Prove that if "A implies B" and "B implies C," then "A implies C."

^{W}**3.13**   Use Exercise 3.12 to help prove that if "A implies B," "B implies C," and "C implies A," then A is equivalent to B and A is equivalent to C.

3.14   Prove that if a and b are rational, then $a + b$ is rational.

^{W}**3.15**   Consider a definition in the form of a statement A, together with three possible alternative definitions, say, B, C, and D. Suppose you want to prove that A is equivalent to each of the three alternatives.

> a. Explain why you can do so by proving that "A implies B," "B implies C," "C implies D," and "D implies A".
>
> b. What is the advantage of the approach in part (a) as opposed to proving separately that A is equivalent to each of statements B, C, and D? (Hint: Count the total number of proofs with the two approaches.)

3.16   Suppose you have already proved the proposition that, "If a and b are nonnegative real numbers, then $(a + b)/2 \geq \sqrt{ab}$."

> a. Explain how you could use this proposition to prove that if a and b are real numbers satisfying the property that $b \geq 2|a|$, then $b \geq \sqrt{b^2 - 4a^2}$. Be careful how you match up notation.
>
> b. Use the foregoing proposition and part (a) to prove that if a and b are real numbers with $a < 0$ and $b \geq 2|a|$, then one of the roots of the equation $ax^2 + bx + a = 0$ is $\leq -b/a$.

^{W}**3.17**   Use the definition of an isosceles triangle to prove that if the right triangle UVW with sides of lengths u and v and hypotenuse of length w satisfies $\sin(U) = \sqrt{u/2v}$, then the triangle UVW is isosceles.

^{B}**3.18**   Use Proposition 1 on page 9 to prove that if the right triangle UVW with sides of lengths u and v and hypotenuse of length w satisfies $\sin(U) = \sqrt{u/2v}$, then the triangle UVW is isosceles.

^{W}**3.19**   Use Proposition 3 on page 28 to prove that if the right triangle UVW with sides of lengths u and v and hypotenuse of length w satisfies $\sin(U) = \sqrt{u/2v}$, then the triangle UVW is isosceles.

4

Quantifiers I:
The Construction Method

In Chapter 3, you saw that a definition is often used to answer a key question. The next four chapters provide several other techniques for formulating and answering a key question that arises when B has a special form.

Two particular forms of statements appear repeatedly throughout all branches of mathematics. They are always identified by certain key words. The first one has the words *there is* (*there are, there exists*); the second one has *for all* (*for each, for every, for any*). These two groups of words are referred to collectively as **quantifiers**, and each one gives rise to its own proof technique. The remainder of this chapter deals with the **existential quantifier** "there is." The **universal quantifier** "for all" is discussed in the next chapter.

4.1 WORKING WITH THE QUANTIFIER "THERE IS"

The quantifier "there is" arises naturally in many mathematical statements. Recall Definition 7 for a rational number as being a real number that can be expressed as the ratio of two integers in which the denominator is not zero. This definition could just as well be written using the quantifier "there are."

> Definition 7. A real number r is **rational** if and only if there are integers p and q with $q \neq 0$ such that $r = p/q$.

Another such example arises from the alternative definition of an even integer, that being an integer that can be expressed as 2 times some integer.

Using a quantifier to express this statement, one obtains:

> Definition 5. An integer n is **even** if and only if there is an integer k such that $n = 2k$.

It is important to observe that the quantifier "there is" allows for the possibility of more than one such object, as is shown in the next definition.

Definition 11 *An integer n is a* **square** *if there is an integer k such that $n = k^2$.*

Note that if a nonzero integer n (say, for example, $n = 9$) is square, then there are two values of k that satisfy $n = k^2$ (in this case, $k = 3$ or -3). More is said in Chapter 11 about the issue of uniqueness—that is, the existence of only one such object.

There are many other instances where an existential quantifier is used, but, from the foregoing examples, you can see that such statements always have the same basic structure. Each time the quantifier "there is," "there are," or "there exists" appears, the statement will have the following **standard form**:

> There is an "object" with a "certain property" such that "something happens."

The words in quotation marks depend on the particular statement under consideration. You must learn to read, to identify, and to write each of the three components. Consider these examples.

1. There is an integer $x > 2$ such that $x^2 - 5x + 6 = 0$.
   Object: integer x.
   Certain property: $x > 2$.
   Something happens: $x^2 - 5x + 6 = 0$.

2. There are real numbers x and y both > 0 such that $2x + 3y = 8$ and $5x - y = 3$.
   Object: real numbers x and y.
   Certain property: $x > 0, y > 0$.
   Something happens: $2x + 3y = 8$ and $5x - y = 3$.

Mathematicians often use the symbol $\exists$ to abbreviate the words "there is" ("there are," and so on) and the symbol $\ni$ for the words "such that" ("for which," and so on). The use of symbols is illustrated in the next example.

3. $\exists$ an angle $t \ni \cos(t) = t$.
   Object: angle t.
   Certain property: none.
   Something happens: $\cos(t) = t$.

Observe that the words "such that" (or equivalent words like "for which") always precede the something that happens. Practice is needed to become fluent at reading and writing these statements.

4.2   HOW TO USE THE CONSTRUCTION METHOD

When proving that "A implies B" is true, suppose you obtain a statement in the forward process that has the quantifier "there is" in the standard form:

A: There is an "object" with a "certain property" such that "something happens."

Thus, you can assume that there is such an object, say, X. This object X together with its certain properties and the something that happens should help you reach the conclusion that B is true. The technique of working with such an object in the forward process is straightforward and is therefore not given a special name.

In contrast, if you encounter the key words "there is" during the backward process, then you must show that

B: There is an "object" with a "certain property" such that "something happens."

One way to do so is to use the **construction method**. The idea is to construct (guess, produce, devise an algorithm to produce, and so on) the desired object. However, you should realize that the construction of the object does not, by itself, constitute the proof. Rather, the proof consists of showing that the object you constructed is in fact the correct one, that is, that the object has the certain property and satisfies the something that happens.

How you actually construct the desired object is not at all clear. Sometimes it is by trial and error; sometimes an algorithm is designed to produce the desired object—it all depends on the particular problem. In almost all cases, the information in statement A is used to help accomplish the task. Indeed, the appearance of the quantifier "there is" in the backward process strongly suggests turning to the forward process to produce the desired object. The construction method is used subtly in the proof of Proposition 2, but another example will clarify the process.

Proposition 4 *If a, b, c, d, e, and f are real numbers such that $ad - bc \neq 0$, then the two equations $ax + by = e$ and $cx + dy = f$ can be solved for the real numbers x and y.*

Analysis of Proof.   On starting the backward process, you should recognize that statement B has the form discussed above, even though the quantifier "there are" does not appear explicitly. To see that this is so, rewrite statement B as follows to contain the quantifier explicitly:

B: There are real numbers x and y such that $ax + by = e$ and $cx + dy = f$.

Statements containing **hidden quantifiers** occur frequently, and you should watch for them.

Proceeding with the construction method, the first step is to identify the objects, the certain property, and the something that happens. In this case:

Objects: real numbers x and y.
Certain property: none.
Something happens: $ax + by = e$ and $cx + dy = f$.

The next step is to construct these real numbers. Turn to the forward process to do so. If you are able to "guess" that $x = (de - bf)/(ad - bc)$ and $y = (af - ce)/(ad - bc)$, then you are fortunate. (Observe that, by guessing these values for x and y, you have used the information in A because the denominators are not 0.) However, recall that constructing these objects does not constitute the proof—you must still show that these objects satisfy the certain property and the something that happens. In this case, that means you must show that, for the foregoing values of x and y, $ax + by = e$ and $cx + dy = f$. In doing this proof, you might write the following:

Noting that $ad - bc \neq 0$, construct
    A1: $x = (de - bf)/(ad - bc)$ and $y = (af - ce)/(ad - bc)$.
It must be shown that, for these values of x and y,
    B1: $ax + by = e$ and $cx + dy = f$.

The remainder of this proof consists of working forward from $A1$ using algebra to show that $B1$ is true. The details are straightforward and are not given here.

While this "guess-and-check" approach is perfectly acceptable for producing the desired x and y, it is not informative as to how these particular values are obtained. A more instructive proof is desirable. For example, these values for x and y are obtained by working backward from the following two equations that you want x and y to satisfy:

$$ax + by = e \tag{4.1}$$
$$cx + dy = f \tag{4.2}$$

On multiplying (4.1) by d and (4.2) by b and then subtracting (4.2) from (4.1), you obtain:

$$(ad - bc)x = de - bf. \tag{4.3}$$

You can then use the information in A to divide (4.3) through by $ad - bc$ because, from the hypothesis, this number is not 0, thus obtaining:

$$x = (de - bf)/(ad - bc).$$

A similar process is used to obtain $y = (af - ce)/(ad - bc)$. In general, it is advisable to work backward from the desired properties to construct the object.

Even with all this added explanation as to how the objects are constructed, the mere construction of the objects does not constitute the proof. You must still show that, for these values of x and y, $ax + by = e$ and $cx + dy = f$.

Proof of Proposition 4.   On multiplying the equation $ax + by = e$ by d, and the equation $cx + dy = f$ by b, and then subtracting the two equations one obtains $(ad - bc)x = (de - bf)$. From the hypothesis, $ad - bc \neq 0$, and so dividing by $ad - bc$ yields $x = (de - bf)/(ad - bc)$. A similar argument shows that $y = (af - ce)/(ad - bc)$. It is not hard to check that, for these values of x and y, $ax + by = e$ and $cx + dy = f$. $\square$

In the foregoing condensed proof, observe that the author glosses over the fact that the constructed objects do satisfy the needed properties. When writing such proofs, be sure to include those details. When reading such proofs, you must fill in those details yourself, as shown next.

4.3   READING A PROOF

The process of reading and understanding a proof is demonstrated with the following proposition.

Proposition 5 *If $m < n$ are consecutive integers and m is even, then 4 divides $m^2 + n^2 - 1$.*

Proof of Proposition 5.   (For reference purposes, each sentence of the proof is written on a separate line.)

> **S1:**   Let $n = m + 1$.
> **S2:**   Then $m^2 + n^2 - 1 = m^2 + (m + 1)^2 - 1 = 2m(m + 1)$.
> **S3:**   Because m is even, there is an integer k such that $m = 2k$.
> **S4:**   Letting $p = k(m + 1)$, it follows that

$$m^2 + n^2 - 1 = 2m(m + 1) = 4k(m + 1) = 4p,$$

and so 4 divides $m^2 + n^2 - 1$.

The proof is now complete. $\square$

Analysis of Proof.   An interpretation of statements $S1$ through $S4$ follows.

Interpretation of S1: *Let $n = m + 1$.*
    The author is working forward from the hypothesis that m and n are consecutive integers.

Interpretation of S2: *Then $m^2 + n^2 - 1 = m^2 + (m + 1)^2 - 1 = 2m(m + 1)$.*
    The author is working forward from $m^2 + n^2 - 1$ by substituting $n = m + 1$ from $S1$ and then using algebra.

Interpretation of S3: *Because m is even, there is an integer k such that $m = 2k$.*

The author is working forward from the hypothesis that m is an even integer by using the definition.

Interpretation of S4: *Letting $p = k(m + 1)$, it follows that ....*
Up to this point, it is unclear why the author derives the statements $S1$, $S2$, and $S3$. The answer lies in $S4$. Specifically, the author is working backward from the conclusion B and asks the key question, "How can I show that an integer (namely, 4) divides another integer (namely, $m^2 + n^2 - 1$)?" The author then uses Definition 1 on page 24 to answer the question, whereby it must be shown that

B1: There is an integer p such that $m^2 + n^2 - 1 = 4p$.

Recognizing the key words "there is" in $B1$, the author uses the construction method to construct the integer p, namely, $p = k(m + 1)$ (which is the reason for statements $S1$, $S2$, and $S3$). Finally, as required by the construction method, the author verifies that this value of p is correct by showing that $m^2 + n^2 - 1 = 4p$.

The following points about reading the foregoing condensed proof of Proposition 5 are in order.

- No mention is made of the techniques being used (the forward-backward and construction methods, and the key questions and answers).

- Several steps are condensed into the single sentence $S4$.

- Although $S1$, $S2$, and $S3$ are true, it is not clear where the author is heading with these statements. When you are not sure of what the author is doing, ask yourself what technique you would use to do the proof. For example, if you work backward yourself from B and recognize that the quantifier "there is" arises, then you will realize that $S1$, $S2$, and $S3$ are part of the construction method.

Summary

The construction method is a technique for dealing with statements in the backward process that have the quantifier "there is" in the standard form:

> There is an "object" with a "certain property" such that "something happens."

To use the construction method:

1. Identify the object, the certain property, and the something that happens in the backward statement containing the quantifier "there is."

2. Turn to the forward process and use the hypothesis together with your creative ability to construct the desired object. To do so, you might also find it useful to work backward from the properties you want the object to satisfy. (The actual construction of the object becomes the new statement it the forward process.)

3. Establish that the object you construct does satisfy the certain property and the something that happens. (The statement that the constructed object satisfies the desired properties becomes the new statement in the backward process.)

Exercises

Note: Solutions to exercises marked with a B are in the back of this book. Solutions to exercises marked with a W are located on the World Wide Web at http://www.wiley.com/college/solow/.

Note: All proofs should contain an analysis of proof and a condensed version. Definitions for all mathematical terms are provided in the glossary at the end of the book.

^{B}**4.1**   For each of the following there-is statements, identify the objects, the certain property, and the something that happens.

a. At a party of n (≥ 2) people, at least two of the people have the same number of friends.

b. The function $f(x)$ has an integer root.

c. There is a point (x, y) with $x \geq 0$ and $y \geq 0$ that lies on the two lines $y = m_1 x + b_1$ and $y = m_2 x + b_2$.

d. Given an angle t, one can find an angle t' between 0 and π whose tangent is larger than that of t.

e. For the two integers a and b, at least one of which is not zero and whose greatest common divisor is c, there are integers m and n such that $am + bn = c$.

4.2   Suppose each of the statements in Exercise 4.1 [except for part (a)] is the conclusion of a proposition. Explain how you would apply the construction method to do the proof. (You need not actually do the proof.)

^{W}**4.3**   Rewrite the following statements using the symbols $\exists$ and $\ni$.

a. A triangle XYZ is isosceles if two of its sides have equal length.

b. A polynomial $p(x)$ of degree n has exactly n complex roots, say, $r_1, \ldots, r_n$.

4.4   Would you use the construction method to prove each of the following propositions? Why or why not? Explain.

  a. The polynomial $x^{71} - 4x^{44} + 11x - 3$ has a real root.

  b. If a and b are integers with $a \neq 0$ for which a does not divide b, then there is no positive integer x such that $ax^2 + bx + b - a = 0$.

  c. If $ABCD$ is a square whose sides have length s, then you can inscribe in $ABCD$ a circle whose area is at least $3s^2/4$.

^{B}**4.5**   Explain why and how the construction method is used in the proof of Proposition 2 on page 25.

^{B}**4.6**   Use the construction method to prove that there is an integer x such that $x^2 - 5x/2 + 3/2 = 0$. Is the constructed object unique?

^{W}**4.7**   Use the construction method to prove that there is a real number x such that $x^2 - 5x/2 + 3/2 = 0$. Is the constructed object unique?

4.8   Prove that if a and b are integers with $a \neq 0$ and x is a positive integer such that $ax^2 + bx + b - a = 0$, then $a|b$.

^{B}**4.9**   Prove that if a, b, and c are integers for which $a|b$ and $b|c$, then $a|c$.

4.10   Prove that if a, b, and c are integers for which $a|(b + c)$ and $a|b$, then $a|c$.

^{W}**4.11**   Prove that if s and t are rational numbers and $t \neq 0$, then s/t is a rational number.

4.12   Consider proving that if a, b, c, d, and e are real numbers for which a certain property P holds, then the intersection of the two sets $S = \{(x,y) : y = ax^2 + bx + c\}$ and $T = \{(x, y) : y = dx + e\}$ is nonempty, that is, $S \cap T \neq \emptyset$. Find a property P that allows you prove this proposition. Then prove the proposition.

^{B}**4.13**   Explain what is wrong with the following condensed proof of the proposition: If a, b, and c are real numbers for which the function $ax^2 + bx + c$ has a rational root, then the function $cx^2 + bx + a$ has a rational root.

> **Proof.** Because the function $ax^2 + bx + c$ has a rational root, there are integers p and q with $q \neq 0$ such that $a(p/q)^2 + b(p/q) + c = 0$. On multiplying through by q^2 and then dividing by p^2, it follows that $x = q/p$ is a rational root of $cx^2 + bx + a$. $\square$

^{W}**4.14**   Do you agree with the following condensed proof that if R, S, and T are sets for which $R \cap S \neq \emptyset$ and $S \cap T \neq \emptyset$, then $R \cap T \neq \emptyset$. Why or why not? Explain.

Proof. It is shown that there is an element in $R \cap T$. Now, because $R \cap S \neq \emptyset$, by definition, there is an element $x \in R \cap S$. Likewise, because $S \cap T \neq \emptyset$, there is an element $x \in S \cap T$. It then follows that $x \in R$ and $x \in T$, so $x \in R \cap T$. $\square$

4.15 Explain what is wrong with the following condensed proof of the proposition that if $m < n$ are consecutive integers and m is even, then 4 divides $m^2 + n^2 - 1$.

Proof. Suppose that $n = m + 1$. The proof follows by noting that for $k = m(m+1)$, $m^2 + n^2 - 1 = 2k$. $\square$

5

Quantifiers II:
The Choose Method

This chapter develops a technique for dealing with statements in the backward process that contain the quantifier "for all." Such statements arise quite naturally in many mathematical areas, one of which is set theory, as you will now see.

5.1 WORKING WITH THE QUANTIFIER "FOR ALL"

A **set** is a collection of items. For example, you can think of the numbers 1, 4, and 7 as a collection of items, and hence they form a set. Each of the individual items is called a **member** or **element of the set** and each member of the set is said to *be in* or *belong to* the set. The set is often denoted by enclosing the list of its members, separated by commas, in braces. Thus, the set consisting of the numbers 1, 4, and 7 is written as follows:

$$\{1, 4, 7\}.$$

To indicate that the number 4 belongs to this set, mathematicians write:

$$4 \in \{1, 4, 7\},$$

where the symbol $\in$ stands for the words "is a member of." Similarly, to indicate that 2 is not a member of $\{1, 4, 7\}$, one would write:

$$2 \notin \{1, 4, 7\}.$$

While it is desirable to make a list of all the elements in a set, sometimes it is impractical to do so because the list is too long. For example, imagine having to write down every integer between 1 and 100,000. When a set has an infinite number of elements (such as the set of real numbers that are greater than or equal to 0) it is impossible to make a complete list, even if you want to. Fortunately there is a way to describe such "large" sets through the use of what is known as **set-builder notation**, which involves using a verbal and mathematical description for the members of the set. With set-builder notation, the set of all real numbers that are greater than or equal to 0 is written as follows:

$$S = \{\text{real numbers } x : x \geq 0\},$$

where the ":" stands for the words "such that." Everything following the ":" is referred to as the **defining property of the set**. The question that one always has to be able to answer is, "How do I know if a particular item belongs to the set?" To answer such a question, you need only check if the item y satisfies the defining property. If so, then y is an element of the set; otherwise, y is not in the set. For the foregoing set S, to see if the real number 3 belongs to S, simply replace x everywhere by 3 and see if the defining property is true. In this case, 3 does belong to S because $3 \geq 0$.

Sometimes part of the defining property appears to the left of the ":" as well as to the right and, when trying to determine if a particular item belongs to such a set, be sure to verify this portion of the defining property, too. For example, if $T = \{\text{real numbers } x \geq 0 : x^2 - x + 2 \geq 0\}$, then -1 does not belong to T even though -1 satisfies the defining property to the right of the ":". The reason is that -1 does not satisfy the defining property to the left of the ":" because -1 is not ≥ 0.

From the point of view of doing proofs, the defining property plays the same role as a definition—the defining property is used to answer the key question, "How can I show that an item belongs to a particular set?" One answer is to check that the item satisfies the defining property.

While discussing sets, observe that it can happen that no item satisfies the defining property. Consider, for example,

$$\{\text{real numbers } x \geq 0 : x^2 + 3x + 2 = 0\}.$$

The only real numbers for which $x^2 + 3x + 2 = 0$ are -1 and -2. Neither of these satisfies the defining property to the left of the ":". Such a set is said to be **empty**, meaning that the set has no members. The special symbol $\emptyset$ is used to denote the empty set.

To motivate the use of the quantifier "for all," observe that it is usually possible to write a set in more than one way, for example, the two sets

$$\begin{aligned} S &= \{\text{real numbers } x : x^2 - 3x + 2 \geq 0\}, \\ T &= \{\text{real numbers } x : 1 \leq x \leq 2\}, \end{aligned}$$

where $1 \leq x \leq 2$ means that $1 \leq x$ and $x \leq 2$. Surely for two sets S and T to be the same, each element of S should appear in T and vice versa. Using the quantifier "for all," a definition can now be made.

Definition 12 *A set S is a **subset** of a set T (written $S \subseteq T$) if and only if for each element $x \in S$, $x \in T$.*

Definition 13 *Two sets S and T are **equal** (written $S = T$) if and only if S is a subset of T and T is a subset of S.*

Like any definition, these are used to answer a key question. Definition 12 answers the question, "How can I show that a set (say, S) is a subset of another set (say, T)?" by requiring you to show that for each element $x \in S$, $x \in T$. How you do so is explained in Section 5.2. Definition 13 answers the key question, "How can I show that two sets (say, S and T) are equal?" by requiring you to show that S is a subset of T and T is a subset of S.

In addition to set theory, there are many other instances where the quantifier "for all" is used, but, from the foregoing example, you can see that such statements appear to have the same consistent structure. When the quantifier "for all," "for each," "for every," or "for any" appears, the statement will have the following **standard form** (which is similar to the one in Chapter 4):

For every "object" with a "certain property," "something happens."

The words in quotation marks depend on the particular statement under consideration, and you must learn to read, write, and identify these three components. Consider these examples.

1. For every angle t, $\sin^2(t) + \cos^2(t) - 1$.
   Object: angle t.
   Certain property: none.
   Something happens: $\sin^2(t) + \cos^2(t) = 1$.

Mathematicians often use the symbol $\forall$ to abbreviate the words "for all" ("for each," and so on). The use of symbols is illustrated in the next example.

2. $\forall$ real numbers $y > 0$, $\exists$ a real number $x \ni 2^x = y$.
   Object: real numbers y.
   Certain property: $y > 0$.
   Something happens: $\exists$ a real number $x \ni 2^x = y$.

Observe that a comma always precedes the something that happens.

Sometimes the quantifier is hidden, for example, the statement "the cosine of an angle strictly between 0 and $\pi/4$ is larger than the sine of the angle" could be phrased equally well as "for every angle t with $0 < t < \pi/4$, $\cos(t) > \sin(t)$." Also, some authors write the quantifier after the something that happens, for example: "$2^n > n^2$, for all integers $n \geq 5$." Practice is needed to become fluent at reading and writing these statements, regardless of how they are presented.

5.2   USING THE CHOOSE METHOD

During the backward process, if you come across a statement having the quantifier "for all" in the standard form:

B: for all "objects" with a "certain property," "something happens,"

then one approach to showing that the statement is true is to make the following list of all of the objects having the certain property:

Objects with the Certain Property

X_1

X_2

X_3

$\vdots$

Then, for each object on the list, you would need to show that the something happens. When the list consists of only a few objects, this might be a reasonable way to proceed. However, when the list is long or infinite, this approach is not practicable. You have already dealt with this type of obstacle in set theory, where the problem is overcome by using set-builder notation to describe the set. Here, the **choose method**, allows you to circumvent the difficulty.

To understand the idea of the choose method consider again the foregoing list of objects, each with the certain property. Suppose you are able to prove that, for object X_1, the something happens. Further, imagine that your proof is such that when you replace X_1 everywhere with X_2, the resulting proof correctly establishes that the something happens for X_2. In this case, you do not need to write a separate proof to establish that the something happens for X_2. You would simply say, "To see that the something happens for X_2, repeat the proof, replacing X_1 everywhere with X_2."

Extending this idea to the remaining objects on the list, the goal of the choose method is to construct a "model proof" for establishing that the something happens for a general object X that has the certain property, in such a way that you could, in theory, repeat the proof for each and every object on the list (see Figure 5.1). If you had such a model proof, then you would not need to check the whole (possibly infinite) list of objects because you would know that you could always do so by a simple substitution in the model proof. In other words, rather than actually proving that the something happens for every object having the certain property, the choose method provides the *capability* of doing so through the use of a model proof. The way in which this model proof is designed is now described.

To understand how the choose method is used, recall that the model proof must establish that the something happens for a general object having the certain property. As such, suppose that you have one of these objects, say, X, but remember, you do not know precisely which one. All you do know

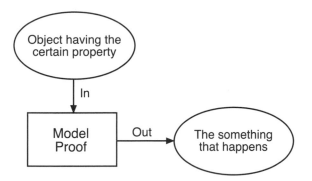

Fig. 5.1   A model proof for the choose method.

is that the particular object X does have the certain property. You must somehow use that property to reach the conclusion that, for this object X, the something happens. This is most easily accomplished by working forward from the certain property and backward from the something that happens. In other words, with the choose method,

- You choose an object with the certain property, which then becomes a new statement in the forward process.

- You must show that, for the chosen object, the something happens, which then becomes the new statement in backward process.

If successful, then you have the capability of repeating the foregoing model proof for any object having the certain property.

An Example of Using the Choose Method

To illustrate how the choose method is used, suppose that, in some proof, you need to show that

B: for all real numbers x with $x^2 - 3x + 2 \le 0$, $1 \le x \le 2$.

The first step is to identify, in this for-all statement, the objects (real numbers x), the certain property ($x^2 - 3x + 2 \le 0$), and the something that happens ($1 \le x \le 2$). To apply the choose method, you choose a real number that has the certain property. In this case, you might write the following:

A1: Let x' be a real number with $(x')^2 - 3x' + 2 \le 0$.

Then, by working forward from $(x')^2 - 3x' + 2 \le 0$, you must reach the conclusion that, for x', the something happens, that is, you must show that

B1: $1 \le x' \le 2$.

Here, the symbol x' is used to distinguish the chosen object from the general object x in B. Notice in $A1$ and in $B1$ that the certain property and the something that happens is written specifically for the chosen object, not for the general one. This is done by replacing the general object (x) everywhere in B with the chosen object (x'). In many condensed proofs, the same symbol is used for both the general object and the chosen one. In such cases, be careful to interpret the symbol correctly. Consider the following example.

Proposition 6 *If S and T are the two sets defined by*

$$
\begin{aligned}
S &= \{real\ numbers\ x : x^2 - 3x + 2 \le 0\} \\
T &= \{real\ numbers\ x : 1 \le x \le 2\},
\end{aligned}
$$

then $S = T$.

Analysis of Proof.   When doing a proof, learn to choose a technique consciously, based on the form of the statements under consideration. In this proposition, the hypothesis A and conclusion B do not contain key words (such as "there is" or "for all"). In the absence of key words, the forward-backward method is a reasonable technique to use. Doing so in this case gives rise to the key question, "How can I show that two sets (namely, S and T) are equal?" Definition 13 provides the answer that you must show that

B1: S is a subset of T and T is a subset of S.

So, first try to establish that

B2: S is a subset of T,

and afterward, that

B3: T is a subset of S.

To show that S is a subset of T (see $B2$), you obtain the key question, "How can I show that a set (namely, S) is a subset of another set (namely, T)?" Using Definition 12 leads to the answer that you must show that

B4: For all elements $x \in S$, $x \in T$.

This new statement, $B4$, contains the quantifier "for all," thus indicating that you should proceed by the choose method. To do so, first identify, in $B4$, the objects (elements x), the certain property (being in S), and the something that happens ($x \in T$).

To apply the choose method to $B4$, you must now choose an object having the certain property and then show that the something happens. In this case that means you should choose

A1: An element $x \in S$.

Using the fact that $x \in S$ (that is, that x satisfies the defining property of S) together with the information in A, you must show that the something happens in $B4$, that is,

B5: $x \in T$.

Note that you do not want to pick one specific element in S, say, $3/2$. Also, note the double use of the symbol x for both the general object in $B4$ and the chosen object in $A1$.

Working backward from $B5$, you should ask the key question, "How can I show that an element (namely, x) belongs to a set (namely, T)?" One answer is to show that x satisfies the defining property of T, that is,

B6: $1 \leq x \leq 2$.

Turning now to the forward process, you can make use of the information in A to show that $1 \leq x \leq 2$ because you have assumed that A is true. However, additional information is available. Recall that, during the backward process, you used the choose method, at which time you chose $x \in S$ (see $A1$). Now is the time to use this fact. Specifically, because $x \in S$, from the defining property of the set S, you have that

A2: $x^2 - 3x + 2 \leq 0$.

Then, by factoring, you obtain

A3: $(x - 2)(x - 1) \leq 0$.

The only way that the product of $x - 2$ and $x - 1$ can be ≤ 0 is for one of the terms to be ≤ 0 and the other ≥ 0. In other words,

A4: Either $x - 2 \geq 0$ and $x - 1 \leq 0$, or else $x - 2 \leq 0$ and $x - 1 \geq 0$.

The first situation can never happen because, if it did, $x \geq 2$ and $x \leq 1$, which is impossible. Thus the second condition must happen, that is,

A5: $x \leq 2$ and $x \geq 1$.

But this is precisely the last statement obtained in the backward process ($B6$), and hence it has been shown successfully that S is a subset of T. Do not forget that you still have to show that T is a subset of S ($B3$) in order to complete the proof that $S = T$. This part is done in Section 5.3.

Proof of Proposition 6.   To show that $S = T$ it is shown that $S \subseteq T$ and $T \subseteq S$. To see that $S \subseteq T$, let $x \in S$ (the use of the word "let" in condensed proofs frequently indicates that the choose method is being invoked). Consequently, $x^2 - 3x + 2 \leq 0$ and so $(x - 2)(x - 1) \leq 0$. This means that either $x - 2 \geq 0$ and $x - 1 \leq 0$ or else $x - 2 \leq 0$ and $x - 1 \geq 0$. The former cannot happen because, if it did, $x \geq 2$ and $x \leq 1$. Hence it must be that $x \leq 2$ and $x \geq 1$, which means that $x \in T$. The proof that T is a subset of S is given in Proposition 7. □

On a final note, whenever you apply the choose method to choose an object with the certain property, you must first be sure that, indeed, there is at least one such object, for if there are none, how can you choose such an object? To illustrate, suppose you are trying to prove that

B: For all real numbers $x \geq 0$ with $x^2 + 3x + 2 = 0$, $x^2 \geq 4$.

According to the choose method, you should choose a real number $x \geq 0$ with the property that $x^2 + 3x + 2 = 0$. However, there are no such numbers (because the only values for x that satisfy the equation are $x = -1$ and $x = -2$, and neither of these is ≥ 0). In this case, you cannot apply the choose method; however, there is no need to do so. The reason is that when there is no object with the certain property, the associated for-all statement is automatically true, as explained in Exercise 5.5.

5.3   READING A PROOF

The process of reading and understanding a proof is now demonstrated.

Proposition 7 *If S and T are the two sets defined by*

$$
\begin{aligned}
S &= \{real\ numbers\ x : x^2 - 3x + 2 \leq 0\} \\
T &= \{real\ numbers\ x : 1 \leq x \leq 2\},
\end{aligned}
$$

then $T \subseteq S$.

Proof of Proposition 7.   (For reference purposes, each sentence of the proof is written on a separate line.)

    S1:   To show that $T \subseteq S$, let $t' \in T$.
    S2:   It is shown that $t' \in S$.
    S3:   Because $t' \in T$, $1 \leq t' \leq 2$, so $t' - 1 \geq 0$ and $t' - 2 \leq 0$.
    S4:   Thus, $t' \in S$ because $(t')^2 - 3t' + 2 = (t' - 1)(t' - 2) \leq 0$.

The proof is now complete. □

Analysis of Proof.   An interpretation of statements $S1$ through $S4$ follows.

Interpretation of S1: *To show that $T \subseteq S$, let $t' \in T$.*
    The author has worked backward from the conclusion B and asked the key question, "How can I show that a set (namely, T) is a subset of another set (namely, S)?" Applying Definition 12 means it must be shown that

B1: for all elements $x \in T$, $x \in S$.

The author then recognizes the key words "for all" in $B1$ and uses the choose method to choose an object with the certain property, as indicated by the words "... let $t' \in T$."

Interpretation of S2: *It is shown that* $t' \in S$.

According to the choose method, it is necessary to show that, for the chosen object, the something happens. This is exactly what the author is saying must be done for the chosen object t'.

Interpretation of S3: *Because* $t' \in T$, $1 \le t' \le 2$, *so* $t'-1 \ge 0$ *and* $t'-2 \le 0$.

The author is working forward from the fact that $t' \in T$, using the defining property of T. Presumably this is being done to show that the something happens for the chosen object, that is, that $t' \in S$.

Interpretation of S4. *Thus,* $t' \in S$ *because* $(t')^2 - 3t' + 2 = (t'-1)(t'-2) \le 0$.

The author is now claiming that $t' \in S$ by showing that t' satisfies the defining property of S. In essence, the author has asked the key question, "How can I show that an element (namely, t') belongs to a set (namely, S)?" and has answered the question by using the defining property of the set. Having shown that $t' \in S$, the choose method, and hence the proof, is now complete.

The following points about reading the condensed proof of Proposition 7 are in order.

- No mention is made of the techniques being used (the forward-backward and choose method, and the key questions and answers). However, in this case, the word "let" indicates that the choose method is used.

- The techniques used in the proof vary as the form of the statement currently under consideration varies. For example, the author starts with the forward-backward method and then changes to the choose method when a backward statement contains the quantifier "for all."

- Several steps are condensed into the single sentence $S1$.

Summary

Use the choose method when the last statement in the backward process contains the quantifier "for all" in the standard form:

For every "object" with a "certain property," "something happens."

To use the choose method, proceed as follows to create a model proof that could, in theory, be repeated for every object with the certain property.

1. Identify the object, the certain property, and the something that happens in the for-all statement.

2. Verify that there is at least one object with the certain property.

3. Choose an object that has the certain property. (Write the fact that the chosen object has the certain property as a new statement in the forward process.)

4. Show that, for this chosen object, the something happens. (Write this objective as the next statement in the backward process.)

Step 4 is accomplished by the forward-backward method. That is, work forward from the fact that the chosen object in Step 3 has the certain property and backward from the fact that this chosen object must be shown to satisfy the something that happens. In so doing, you can use the assumption that the hypothesis A, or any other statement in the forward process, is true.

Exercises

Note: Solutions to exercises marked with a B are in the back of this book. Solutions to exercises marked with a W are located on the World Wide Web at http://www.wiley.com/college/solow/.

Note: All proofs should contain an analysis of proof and a condensed version. Definitions for all mathematical terms are provided in the glossary at the end of the book.

^{B}**5.1**   For each of the following definitions, identify the objects, the certain property, and the something that happens in the associated for-all statements.

a. The real number x^* is a **maximum** of the function f if and only if for every real number x, $f(x) \leq f(x^*)$.

b. Suppose that f and g are functions of one variable. Then $g \geq f$ **on the set S** of real numbers if and only if for every element $x \in S$, $g(x) \geq f(x)$.

c. A real number u is an **upper bound** for a set S of real numbers if and only if for all $x \in S$, $x \leq u$.

d. The set C of real numbers is a **convex set** if and only if for all elements $x, y \in C$ and for every real number t with $0 \leq t \leq 1$, $tx + (1 - t)y \in C$.

e. The function f of one variable is a **convex function** if and only if for all real numbers x and y and for all real numbers t with $0 \leq t \leq 1$, it follows that $f(tx + (1 - t)y) \leq tf(x) + (1 - t)f(y)$.

5.2   Suppose you are trying to prove that, "If S and T are the sets defined by $S = \{(x, y) : x^2 + y^2 \leq 16\}$ and $T = \{(x, y) : 3x^2 + 2y^2 \leq 125\}$, then for every element $(x, y) \in S$, $(x, y) \in T$." Which of the following constitutes a correct application of the choose method? For those that are incorrect, explain what is wrong.

a.   Choose
    $A1$: real numbers x' and y'.
    It must be shown that
    $B1 : (x', y') \in T$.

b.   Choose
    $A1$: real numbers x' and y' with $(x', y') \in S$.
    It must be shown that
    $B1 : (x', y') \in T$.

c.   Choose
    $A1$: real numbers x' and y' with $(x', y') \in T$.
    It must be shown that
    $B1 : (x', y') \in S$.

d.   Choose
    $A1$: real numbers x' and y', say, 1 and 2, with
        $x^2 + y^2 = 5 \leq 16$ and therefore $(x', y') \subset S$.
    It must be shown that
    $B1 : 3x^2 + 2y^2 \leq 125$ and therefore that $(x', y') \in T$.

e.   Choose
    $A1$: real numbers x and y with $(x, y) \in S$.
    It must be shown that
    $B1 : (x, y) \in T$.

[B]**5.3**   For each of the parts in Exercise 5.1, describe how you would apply the choose method to show that the for-all statement is true. Use a different symbol to distinguish the chosen object from the general object. For instance, for Exercise 5.1(a), to show that x^* is the maximum of the function f, you would choose

   A1: a real number, say, x',

for which it must then be shown that

   B1: $f(x') \leq f(x^*)$.

5.4   Would you use the chose method to prove that for all real numbers a, b, and c, if $4ac \leq b^2$, then $ax^2 + bx + c$ has real roots? Why or why not? Explain.

[B]**5.5**   Consider the problem of showing that, "For every object X with a certain property, something happens." Discuss why the approach of the choose method is the same as that of using the forward-backward method to show that, "If X is an object with the certain property, then the something happens." How are the two statements in quotation marks related? Use this relationship to explain why a for-all statement in which there is no object with the certain property is necessarily true. (Hint: What does such a for-all statement mean about the hypothesis of the associated if/then statement?)

5.6   Using the results of Exercise 5.5, reword the following for-all statements in an equivalent "If ... then ..." form.

a. For every prime number p, $p + 7$ is composite.

b. For all sets A, B, and C with the property that $A \subseteq B$ and $B \subseteq C$, it follows that $A \subseteq C$.

c. For all integers p and q with $q \neq 0$, p/q is rational.

^{W}**5.7**   Reword the following statements in standard form using the appropriate symbols $\forall$, $\exists$, $\ni$, as necessary.

a. Some mountain is taller than every other mountain.

b. If t is an angle, then $\sin(2t) = 2\sin(t)\cos(t)$. (Hint: Use Exercise 5.5.)

c. The square root of the product of any two nonnegative real numbers p and q is not less than their sum divided by 2.

d. If x and y are real numbers such that $x < y$, then there is a rational number r such that $x < r < y$. (Hint: Use Exercise 5.5.)

^{B}**5.8**   For the proposition and condensed proof given below, explain where (that is, in which sentence), why, and how the choose method is used. Identify any mistakes you encounter. [Refer to the definition in Exercise 5.1(a).]

> **Proposition.** If a, b, and c are real numbers for which $a < 0$, then $x^* = -b/(2a)$ is a maximum of $f(x) = ax^2 + bx + c$.

> **Proof.** Let x be a real number. If $x^* \geq x$, then $x^* - x \geq 0$ and $a(x^* + x) + b \geq 0$. So, $(x^* - x)[a(x^* + x) + b] \geq 0$. On multiplying the term $x^* - x$ through, rearranging terms, and adding c to both sides, one obtains that $a(x^*)^2 + bx^* + c \geq ax^2 + bx + c$. A similar argument applies when $x^* < x$. $\square$

5.9   For the proposition and condensed proof given below, explain where (that is, in which sentence), why, and how the choose method is used. Identify any mistakes you encounter.

Proposition. If

$$R = \{\text{real numbers } x : x^2 - x \geq 0\},$$
$$S = \{\text{real numbers } x : -(x-1)(x-3) \leq 0\}, \quad \text{and}$$
$$T = \{\text{real numbers } x : x \leq 1\},$$

then $R \cap S \subseteq T$.

Proof. To reach the conclusion, it will be shown that for all $x \in R \cap S$, $x \in T$. To that end, let $x \in R \cap S$. Because $x \in R$, $x^2 - x = x(x-1) \geq 0$. Likewise, because $x \in S$, $-(x-1)(x-3) \leq 0$. Combining these two yields that $x \leq 1$ and so $x \in T$. $\square$

^{B}**5.10** Write an analysis that corresponds to the condensed proof given below. Indicate which techniques are used and how they are applied. Fill in the details of any missing steps where appropriate.

Proposition. If m and b are real numbers with $m > 0$, and f is the function defined by $f(x) = mx + b$, then for all real numbers x and y with $x < y$, $f(x) < f(y)$.

Proof. Let x and y be real numbers with $x < y$. Then because $m > 0$, $mx < my$. On adding b to both sides, it follows that $f(x) < f(y)$, and so the proof is complete. $\square$

^{W}**5.11** Write an analysis that corresponds to the condensed proof given below. Indicate which techniques are used and how they are applied. Fill in the details of any missing steps where appropriate.

Proposition. If $S = \{\text{real numbers } x : x(x-3) \leq 0\}$ and $T = \{\text{real numbers } x : x \geq 3\}$, then every element of T is an upper bound for the set S [see the definition in Exercise 5.1(c)].

Proof. Let t be an element of T. It will be shown that t is an upper bound for S. To that end, let x be an element of S. Consequently, $x(x-3) \leq 0$. Therefore, it must be that $x \geq 0$ and $x - 3 \leq 0$. But then $x \leq 3$, and because t belongs to T, $t \geq 3$, and so $x \leq t$. $\square$

5.12 Write an analysis that corresponds to the condensed proof given below. Indicate which techniques are used and how they are applied. Fill in the details of any missing steps where appropriate.

Proposition. If p is a positive integer, then for all nonzero integers q and r having the same sign and for which $q < r$, $p/q > p/r$.

Proof. Because q and r have the same sign, $p/(qr) > 0$. Now $r > q$, so multiplying both sides by $p/(qr)$ results in $rp/(qr) > qp/(qr)$. It now follows that $p/q > p/r$, and so the proof is complete. □

^{B}**5.13**   Prove that if $f(x) = (x - 1)^2$ and $g(x) = x + 1$, then $g \geq f$ on the set $S = \{$real numbers $x : 0 \leq x \leq 3\}$. [See the definition in Exercise 5.1(b).]

5.14   Prove that for all real numbers a and b, at least one of which is not 0, $a^2 + ab + b^2 > 0$. (Hint: Use the fact that $a^2 + b^2 > (a^2 + b^2)/2$.)

5.15   A function f of one variable is **strictly increasing** if and only if for all real numbers x and y with $x < y$, $f(x) < f(y)$. Use the result in Exercise 5.14 to prove that the function $f(x) = x^3$ is strictly increasing.

^{B}**5.16**   Prove that if m and b are real numbers and f is the function defined by $f(x) = mx + b$, then f is convex. [See the definition in Exercise 5.1(e).]

^{W}**5.17**   Prove that if a and b are real numbers, then the set $C = \{$real numbers $x : ax \leq b\}$ is a convex set. [See the definition in Exercise 5.1(d).]

6

Quantifiers III: Specialization

The previous two chapters illustrate how to proceed when a quantifier appears in the statement B. A method is introduced in this chapter for working forward from a statement that contains the universal quantifier "for all."

6.1 HOW TO USE SPECIALIZATION

When the statement A contains the quantifier "for all" in the standard form:

> **A:** For all "objects" with a "certain property,"
> "something happens,"

one typical method emerges for working forward from A—**specialization**. In general terms, specialization works as follows. As a result of assuming A is true, you know that, for all objects with the certain property, something happens. If, at some point, you were to come across one of these objects that does have the certain property, then you can use the information in A by being able to conclude that, for this particular object, the something does indeed happen. That fact should help you to conclude that B is true. In other words, you will have specialized the statement A to one particular object having the certain property.

To illustrate the idea of specialization in a more tangible way, suppose you know that

> **A:** All Fords made in 2000 get good gas mileage.

In this statement, you can identify the following three items:

Objects: Fords.
Certain property: made in 2000.
Something happens: get good gas mileage.

Suppose that you are interested in buying a car that gets good gas mileage, so your objective is

B: To buy a car that gets good gas mileage.

You can work forward from the information in A by specialization to establish B as follows. Suppose you are in a dealer's lot one day and you see a particular Ford. Looking more carefully at the car, you verify that the Ford was made in 2000. Recalling that statement A above is assumed to be true, you can use this information to conclude that

A1: This particular 2000 Ford gets good gas mileage.

In other words, you have specialized the for-all statement in A to one particular object.

If you analyze the example above in detail, you can identify the following steps associated with applying specialization to a forward statement of the form:

A: For all "objects" with a "certain property,"
     "something happens."

Steps for Using Specialization

1. Identify, in the for-all statement, the objects, the certain property, and the something that happens.

2. Look for one particular object to apply specialization to. (This object often arises in the backward process.)

3. Verify that this particular object does have the certain property specified in the for-all statement.

4. Conclude, by writing a new statement in the forward process, that the something happens for this particular object.

The following example demonstrates the proper use of specialization in doing mathematical proofs.

Definition 14 *A real number u is an* **upper bound** *for a set of real numbers T if and only if for all elements $t \in T$, $t \le u$.*

Proposition 8 *If R is a subset of a set S of real numbers and u is an upper bound for S, then u is an upper bound for R.*

Analysis of Proof. The forward-backward method gives rise to the key question, "How can I show that a real number (namely, u) is an upper bound for a set of real numbers (namely, R)?" Definition 14 is used to answer the question. Thus it must be shown that

B1: For all elements $r \in R$, $r \leq u$.

The appearance of the quantifier "for all" in the backward process suggests proceeding with the choose method, whereby one chooses

A1: An element, say, $r \in R$.

for which it must be shown that

B2: $r \leq u$.

(Note that the same symbol r is used for the chosen object in $A1$ as for the general object in the for-all statement in $B1$.)

Turning now to the forward process, you will see how specialization is used to reach the conclusion that $r \leq u$. From the hypothesis that R is a subset of S, and by Definition 12 on page 46, you know that

A2: For each element $x \in R$, $x \in S$.

Recognizing the key words "for all" in the forward process, you should consider using specialization. According to the discussion preceding Proposition 8, the first step in doing so is to identify, in $A2$, the object (element x), the certain property (being in the set R), and the something that happens ($x \in S$). Next, you must look for one particular object with which to specialize. In the backward process you chose the particular element r (see $A1$). The third step of specialization requires that you verify that the particular object (r) has the certain property in $A2$, namely, being in the set R. In fact, the particular object r does have this property because r is chosen to be in the set R (see $A1$). The final step of specialization is to conclude, by writing a new statement in the forward process, that, for the particular object r, the something in $A2$ happens. In this case, specialization allows you to conclude that

A3: $r \in S$.

The proof is not yet complete because the last statement in the backward process ($B2$) has not yet been reached in the forward process. To do so, continue to work forward. For example, from the hypothesis, you know that u is an upper bound for S. By Definition 14 this means that

A4: For every element $s \in S$, $s \leq u$.

Once again, the appearance of the quantifier "for every" in the forward process suggests using specialization. Accordingly, identify, in $A4$, the object (element s), the certain property (being in the set S), and the something that happens

($s \le u$). Now look for one particular object with which to apply specialization. The same element r chosen in $A1$ serves the purpose. Next, verify that this particular r satisfies the certain property in $A4$, namely, of being in the set S. Indeed, $r \in S$, as stated in $A3$. Finally, conclude that, for this particular object r, the something in $A4$ happens, so

   A5: $r \le u$.

The proof is now complete because $A5$ is the last statement obtained in the backward process (see $B2$).

   In the condensed proof that follows, note the lack of reference to the forward-backward, choose, and specialization methods.

Proof of Proposition 8.   To show that u is an upper bound for R, let $r \in R$ (the word "let" here indicates that the choose method is used). By hypothesis, $R \subseteq S$ and so $r \in S$ (here is where specialization is used). Furthermore, by hypothesis, u is an upper bound for S, thus, every element in S is $\le u$. In particular, $r \in S$, so $r \le u$ (again specialization is used). $\square$

   When using specialization, be careful to keep your notation and symbols in order. Doing so involves a correct "matching up of notation," similar to what you learned in Chapter 3 on using definitions. To illustrate, suppose you are going to apply specialization to a statement of the form:

   A: For all objects X with a certain property,
         something happens.

After finding a particular object, say, Y, with which to specialize, it is necessary to verify that Y satisfies the certain property in A. To do so, replace X with Y everywhere in the certain property in A and see if the resulting condition is true. Similarly, when concluding that the particular object Y satisfies the something that happens in A, again replace X everywhere with Y in the something that happens to obtain the correct statement in the forward process. (This is done when writing statements $A3$ and $A5$ in the foregoing proof of Proposition 8.) Be careful of overlapping notation, for example, when the particular object you have identified has precisely the same symbol as the one in the for-all statement you are specializing.

6.2   READING A PROOF

The process of reading and understanding a proof is demonstrated with the following proposition.

Definition 15 *A real number* u *is a* **least upper bound** *for a set* S *of real numbers if and only if (1)* u *is an upper bound for* S *and (2) for every upper bound* v *for* S, $u \le v$.

Proposition 9 *If v^* and w^* are least upper bounds for a set T, then $v^* = w^*$.*

Proof of Proposition 9.   (For reference purposes, each sentence of the proof is written on a separate line.)

S1:   From the hypothesis, both v^* and w^* are upper bounds for T.
S2:   Because v^* is a least upper bound for T, $v^* \leq u$, for any upper bound u for T.
S3:   In particular, w^* is an upper bound for T, so $v* \leq w^*$.
S4:   Similarly, w^* is a least upper bound for T and because v^* is an upper bound for T, $w^* \leq v^*$.
S5:   It now follows that $v^* = w^*$.

The proof is now complete. □

Analysis of Proof.   An interpretation of statements $S1$ through $S5$ follows.

Interpretation of S1: *From the hypothesis, both v^* and w^* are upper bounds for T.*
    The author is working forward from the hypothesis using part (1) of the definition of a least upper bound to claim that both v^* and w^* are upper bounds for T.
    When reading a proof, it is advisable to determine where the author is heading. To do so, work backward from B yourself. In this case, you are led to the key question, "How can I show that two real numbers (namely, v^* and w^*) are equal?" Read forward in the proof to see how the author answers this question. From $S3$ and $S4$, the answer in this case is to show that

B1: $v^* \leq w^*$ and $w^* \leq v^*$.

Interpretation of S2: *Because v^* is a least upper bound for T, $v^* \leq u$, for any upper bound u for T.*
    The author is continuing to work forward by stating part (2) of the definition of a least upper bound applied to v^*, that is,

A1: For every upper bound u for T, $v^* \leq u$.

Interpretation of S3: *In particular, w^* is an upper bound for T, so $v^* \leq w^*$.*
    It is here that specialization is applied to the for-all statement in $A1$. In particular, $A1$ is specialized to the value $u = w^*$, which is an upper bound for T (see $S1$). The result of specialization is

A2: $v^* \leq w^*$,

which is just what the author is claiming in $S3$.

Interpretation of S4: *Similarly, w^* is a least upper bound for T and because v^* is an upper bound for T, $w^* \leq v^*$.*

The author is using the same analysis as in $S3$, but this time, applied to the least upper bound w^* and the upper bound v^* for T. The result of this specialization is that

A3: $w^* \leq v^*$.

Interpretation of S5: *It now follows that $v^* = w^*$, and so the proof is complete.*

The author is working forward by combining $v^* \leq w^*$ from $A2$ and $w^* \leq v^*$ from $A3$ to claim correctly that $v^* = w^*$. Finally, the author states that the proof is complete, which is true because the conclusion B has been established.

Summary

You now have various techniques for dealing with quantifiers that can appear in either A or B. As always, let the form of the statement guide you. When B contains the quantifier "there is," the construction method is used to produce the desired object. The choose method is associated with the quantifier "for all" in the backward process. Finally, if the quantifier "for all" appears in the forward process, use specialization. To do so, follow these steps:

1. Identify, in the for-all statement, the objects, the certain property, and the something that happens.

2. Look for one particular object to apply specialization to. (This object often arises from the backward process, especially when the choose method is used.)

3. Verify that this particular object does have the certain property specified in the for-all statement.

4. Conclude, by writing a new statement in the forward process, that the something happens for this one particular object.

It is common to confuse the choose method with the specialization method. Use the choose method when you encounter the key words "for all" in the backward process; use specialization when the key words "for all" arise in the forward process. Another way to say this is to use the choose method when you want to *show that* "for all objects with a certain property, something happens"; use specialization when you *know that* "for all objects with a certain property, something happens."

All of the statements thus far have contained only one quantifier. In the next chapter you will learn what to do when those statements contain more than one quantifier.

Exercises

Note: Solutions to exercises marked with a B are in the back of this book. Solutions to exercises marked with a W are located on the World Wide Web at http://www.wiley.com/college/solow/.

Note: All proofs should contain an analysis of proof and a condensed version. Definitions for all mathematical terms are provided in the glossary at the end of the book.

^{W}**6.1** Suppose you want to specialize the statement that "for every object with a certain property, something happens" to the particular object Y. Explain why you need to show that Y has the certain property before you can do so.

^{B}**6.2** Suppose you are working backward and want to show that, for a particular object Y with a certain property P, S happens. When and how can you reach this conclusion by working forward from the statement that for every object X with a certain property Q, T happens. State your answer in terms of the objects, the certain properties, and the somethings that happen.

^{B}**6.3** For each definition in Exercise 5.1 on page 54, explain how you would work forward from the associated for-all statement. For example, to work forward from the for-all statement in Exercise 5.1(b), (1) look for a specific element, say, y, with which to apply specialization, (2) show that $y \in S$, and (3) conclude that $g(y) \geq f(y)$ as a new statement in the forward process.

6.4 For each of the following for-all statements, what properties must the given object satisfy so that you can apply specialization? Given that the object does satisfy those properties, what can you conclude about the object?

a. Statement: For all prime numbers p, $p + 7$ is composite.
    Given object: An integer m.

b. Statement: For all elements $x > 0$ in a set S of real numbers, x is a root of the polynomial $p(x)$.
    Given object: A real number y.

c. Statement: Every triangle ABC with sides of length $a = \overline{BC}$, $b = \overline{CA}$, and $c = \overline{AB}$, satisfies $c^2 = a^2 + b^2 - 2ab\cos(C)$.
    Given object: The isosceles right triangle ABC whose legs $a = \overline{BC}$ and $b = \overline{CA}$ are both equal to m.

d. Statement: For all pairs of equilateral triangles ABC and DEF, if one side of triangle ABC is parallel to one side of triangle DEF, then the other two sides of triangle ABC are parallel to the corresponding sides of triangle DEF.
    Given object: The triangle CDE whose side DE is parallel to side DA of triangle FDA.

^{B}**6.5**   To what specific object could you specialize each of the following for-all statements so that the result of specialization leads to the desired conclusion? Verify that the object to which you are applying specialization satisfies the certain property in the for-all statement so that you can apply specialization.

a. For-all statement: For all angles α and β, $\sin(\alpha + \beta) = \sin(\alpha)\cos(\beta) + \cos(\alpha)\sin(\beta)$.

Desired conclusion: For a particular angle X, $\sin(2X) = 2\sin(X)\cos(X)$.

b. For-all statement: For any sets S and T, $(S \cup T)^c = S^c \cap T^c$ (where X^c is the complement of the set X).

Desired conclusion: For two sets A and B, $(A \cap B)^c = A^c \cup B^c$.

6.6   To what specific object could you specialize each of the following for-all statements so that the result of specialization leads to the desired conclusion? Verify that the object to which you are applying specialization satisfies the certain property in the for-all statement so that you can apply specialization.

a. For-all statement: f is a function of one variable such that for all real numbers x, y, and t with $0 \le t \le 1$, $f(tx + (1-t)y) \le tf(x) + (1-t)f(y)$.

Desired conclusion: the function f satisfies $f(1/2) \le (f(0) + f(1))/2$.

b. For-all statement: for all real numbers c and d for which $c^2 \ge d^2$, $\sqrt{c^2 - d^2} \le c$.

Desired conclusion: for the real numbers a, $b \ge 0$, $\sqrt{ab} \le (a+b)/2$.

^{B}**6.7**   For sets R, S, and T, prove that if $R \subseteq S$ and $S \subseteq T$, then $R \subseteq T$.

6.8   For real numbers u and v, prove that if u is an upper bound for a set S of real numbers and $u \le v$, then v is an upper bound for S.

^{B}**6.9**   Prove that if S and T are convex sets (see Exercise 5.1(d) on page 54), then $S \cap T$ is a convex set.

6.10   For functions f and g of one variable, prove that if $g \ge f$ on the set of real numbers and x^* is a maximum of g (see Exercise 5.1(a) and (b) on page 54), then for every real number x, $f(x) \le g(x^*)$.

^{W}**6.11**   Prove that if f is a convex function of one variable (see Exercise 5.1(e) on page 54), then for all real numbers $s \ge 0$, the function sf is convex [where the value of the function sf at any point x is $sf(x)$].

6.12   Suppose that a, b, and c are real numbers. Prove that if $x^* = -b/(2a)$ is a maximum of the function $f(x) = ax^2 + bx + c$ [see Exercise 5.1(a) on page 54)], then $a \le 0$. (Hint: Specialize x to a value of $x^* + \epsilon$, where $\epsilon > 0$.)

^{B}**6.13**   Write an analysis of proof that corresponds to the condensed proof given below. Indicate which techniques are used and how they are applied. Fill in the details of any missing steps, where appropriate.

> **Proposition.** If R is subset of a set S of real numbers and if f and g are functions for which $g \geq f$ on S (see the definition in Exercise 5.1(b) on page 54), then $g \geq f$ on R.

> **Proof.** To show that $g \geq f$ on R, let $x \in R$. Because R is a subset of S, every element $r \in R$ is in S. In particular, $x \in R$, so $x \in S$. Also, because $g \geq f$ on S, it follows that for every element $s \in S$, $g(s) \geq f(s)$. In particular, $x \in S$, so $g(x) \geq f(x)$. □

6.14   Write an analysis of proof that corresponds to the condensed proof given below. Indicate which techniques are used and how they are applied. Fill in the details of any missing steps, where appropriate.

> **Proposition.** If f and g are convex functions (see the definition in Exercise 5.1(e) on page 54), then $f + g$ is a convex function.

> **Proof.** To see that $f + g$ is convex, let x, y, and t be real numbers with $0 \leq t \leq 1$. Then, because f is convex, it follows that

$$f(tx + (1 - t)y) \leq tf(x) + (1 - t)f(y) \quad \text{(i)}$$

> Likewise, because g is convex,

$$g(tx + (1 - t)y) \leq tg(x) + (1 - t)g(y) \quad \text{(ii)}$$

> Adding (i) and (ii) yields that

$$f(tx+(1-t)y)+g(tx+(1-t)y) \leq t[f(x)+g(x)]+(1-t)[f(y)+g(y)].$$

> Thus, $f + g$ is convex and the proof is complete. □

^{B}**6.15**   What, if anything, is wrong with the following proof?

> **Proposition.** If R is a nonempty subset of a set S of real numbers and R is convex (see Exercise 5.1(d) on page 54), then S is convex.

> **Proof.** To show that S is a convex set, let x, $y \in S$, and t be a real number with $0 \leq t \leq 1$. Because R is a convex set, by definition, for any two elements u and v in R, and for any real

number s with $0 \leq s \leq 1$, $su + (1 - s)v \in R$. In particular, for the specific elements x and y, and for the real number t, it follows that $tx + (1 - t)y \in R$. Because $R \subseteq S$, it follows that $tx + (1 - t)y \in S$ and so S is convex. □

6.16   What, if anything, is wrong with the following proof?

Proposition. If a, b, and c are real numbers with $a < 0$ and x^* is a maximum of $f(x) = ax^2 + bx + c$ (see Exercise 5.1(a) on page 54), then for every real number $\epsilon > 0$, $\epsilon \leq (2ax^* + b)/a$.

Proof. Let $\epsilon > 0$. It will be shown that $\epsilon \leq (2ax^* + b)/a$. Because x^* is a maximum of f, by definition, for every real number x, $f(x^*) \geq f(x)$. In particular, for $x = x^* - \epsilon$, you have that

$$
\begin{aligned}
a(x^*)^2 + bx^* + c &\leq a(x^* - \epsilon)^2 + b(x^* - \epsilon) + c \\
&= a(x^*)^2 + bx^* + c - (b + 2ax^*)\epsilon + a\epsilon^2.
\end{aligned}
$$

Subtracting $a(x^*)^2 + bx^* + c$ from both sides and dividing by $\epsilon > 0$, it follows that

$$
a\epsilon - (b + 2ax^*) \geq 0.
$$

The desired result that $\epsilon \leq (2ax^* + b)/a$ follows by adding $b + 2ax^*$ to both sides of the foregoing inequality and dividing by $a < 0$. □

7

Quantifiers IV: Nested Quantifiers

The statements in the previous three chapters contain only one quantifier—either "there is" or "for all." You will now learn what to do when statements contain more than one quantifier, that is, when there are **nested quantifiers**.

7.1 UNDERSTANDING STATEMENTS WITH NESTED QUANTIFIERS

The use of nested quantifiers is illustrated in the following statement containing both "for all" and "there is":

> **S1:** For all real numbers x with $0 \le x \le 1$, there is a real number y with $-1 \le y \le 1$ such that $x + y^2 = 1$.

When reading, writing, or processing such statements, always work from left to right. For the foregoing statement $S1$, the first quantifier encountered from the left is "for all." For that quantifier, identify the following components:

Object:	real number x.
Certain property:	$0 \le x \le 1$.
Something happens:	there is a real number y with $-1 \le y \le 1$ such that $x + y^2 = 1$.

The something that happens in this case contains the nested quantifier "there is," for which you can then identify the following components:

Object:	real number y.
Certain property:	$-1 \leq y \leq 1$.
Something happens:	$x + y^2 = 1$.

As another example of nested quantifiers, suppose that T is a set of real numbers and consider the statement:

S2: There is a real number $M > 0$ such that for all elements $x \in T$, $x < M$.

In $S2$, the quantifier "there is" is the first one encountered when reading from the left. Associated with that quantifier are the following three components:

Object:	real number M.
Certain property:	$M > 0$.
Something happens:	for all elements $x \in T$, $x < M$.

The something that happens in this case contains the nested quantifier "for all," for which you can identify the following components:

Object:	element x.
Certain property:	$x \in T$.
Something happens:	$x < M$.

It is important to realize that the order in which the quantifiers appear is critical to the meaning of the statement. For example, compare the foregoing statement $S2$ with the following statement:

S3: For all real numbers $M > 0$, there is an element $x \in T$ such that $x < M$.

$S3$ states that for each positive real number M, you can find a (possibly different) element $x \in T$ for which $x < M$. Note that the element x may depend on the value of M, that is, if the value of M is changed, the value of x may change. In contrast, $S2$ says that there is a positive real number M such that, no matter which element x you choose in T, $x < M$. It should be clear that $S2$ and $S3$ are not the same.

Statements can have any number of quantifiers, as illustrated in the following example containing three nested quantifiers (in which f is a function of one variable):

S4: For every real number $\epsilon > 0$, there is a real number $\delta > 0$ such that for all real numbers x and y with $|x - y| < \delta$, $|f(x) - f(y)| < \epsilon$.

Applying the principle of working from left to right, associated with the first quantifier "for all" in $S4$ you should identify:

Object:	real number ϵ.
Certain property:	$\epsilon > 0$.
Something happens:	there is a real number $\delta > 0$ such that for all real numbers x and y with $|x - y| < \delta$, $|f(x) - f(y)| < \epsilon$.

The something that happens in this case contains the nested quantifiers "there is" and "for all." Working from left to right once again, identify the three components associated with the quantifier "there is":

Object:	real number δ.
Certain property:	$\delta > 0$.
Something happens:	for all real numbers x and y with $|x-y| < \delta$, $|f(x) - f(y)| < \epsilon$.

The three components of the last nested quantifier "for all" in the foregoing something that happens are:

Objects:	real numbers x and y.
Certain property:	$|x - y| < \delta$.
Something happens:	$|f(x) - f(y)| < \epsilon$.

7.2 USING PROOF TECHNIQUES WITH NESTED QUANTIFIERS

When a statement in the forward or backward process contains nested quantifiers, apply appropriate techniques (construction, choose, and specialization) based on the order of the quantifiers from left to right in the statement. To illustrate, suppose you want to show that the foregoing statement $S2$ is true, that is, you want to show that

> **B:** There is a real number $M > 0$ such that for all elements $x \in T$, $x < M$.

Because B is in the backward process and the first quantifier encountered from the left is "there is," the construction method is the correct proof technique to use first. Thus, you should turn to the forward process in an attempt to construct a real number $M > 0$. Suppose you have done so. According to the construction method, you must show that the value of M you constructed satisfies the something that happens, that is, you must show that

> **B1:** For all elements $x \in T$, $x < M$.

When trying to show that B1 is true, you should apply the choose method because of the appearance of the quantifier "for all" in the backward process. In this case, you would choose

A1: An element $x \in T$,

for which it must be shown that

B2: $x < M$.

Applying appropriate techniques to nested quantifiers is illustrated again in the proof of the following proposition.

Definition 16 *A function f of one variable is* **onto** *if and only if for every real number y, there is a real number x such that $f(x) = y$.*

Proposition 10 *If m and b are real numbers with $m \neq 0$, then the function $f(x) = mx + b$ is onto.*

Analysis of Proof.   The forward-backward method is used to begin the proof because the hypothesis A and the conclusion B do not contain key words (such as "for all" or "there is"). Working backward, you are led to the key question, "How can I show that a function (namely, $f(x) = mx + b$) is onto?" Applying Definition 16 to the specific function $f(x) = mx + b$, you must show that

B1: For every real number y, there is a real number x such that
$mx + b = y$.

The statement $B1$ contains nested quantifiers. In deciding which proof technique to apply next, observe that the quantifier "for all" is the first one encountered from the left. Thus, use the choose method (because the key words "for all" appear in the backward process). Accordingly, you should choose

A1: A real number y,

for which it must be shown that

B2: There is a real number x such that $mx + b = y$.

Recognizing the key words "there is" in $B2$, you should now proceed with the construction method. To that end, turn to the forward process in an attempt to construct the desired real number x.

Looking at the fact that you want $mx + b = y$ and observing that $m \neq 0$ by the hypothesis, you might produce this statement:

A2: Construct the real number $x = (y - b)/m$.

Recall that, with the construction method, you must show that the object you constructed satisfies the certain property and the something that happens (in $B2$). Thus you must show that

B3: $mx + b = y$.

But, from $A2$,

A3: $mx + b = m[(y - b)/m] + b = (y - b) + b = y,$

and so the proof is complete.

In the condensed proof that follows, observe that the names of the techniques are omitted. Note also that several steps in the foregoing analysis-of-proof are combined into a single statement.

Proof of Proposition 10. To show that f is onto, let y be a real number. (The word "let" here indicates that the choose method is used.) Because, by hypothesis, $m \neq 0$, let $x = (y - b)/m$. (Here, the word "let" indicates that the construction method is used.) It is easy to see that $f(x) = mx + b = y$, and so the proof is complete. □

7.3   READING A PROOF

The process of reading and understanding a proof is demonstrated with the following proposition.

Proposition 11 *If a, b, and c are real numbers with $a < 0$, then there is a real number y such that for every real number x, $ax^2 + bx + c < y$.*

Proof of Proposition 11.   (For reference purposes, each sentence of the proof is written on a separate line.)

S1:   Let $y = \frac{4ac - b^2}{4a}$.

S2:   For a real number x, you have that

$$ax^2 + bx + c = a \left(x + \frac{b}{2a} \right)^2 + \frac{4ac - b^2}{4a}.$$

S3:   Because $a < 0$, it follows that

$$ax^2 + bx + c \leq \frac{4ac - b^2}{4a} = y.$$

The proof is now complete. □

Analysis of Proof.   An interpretation of statements $S1$ through $S3$ follows.

Interpretation of S1: *Let $y = \frac{4ac - b^2}{4a}$.*
The author recognizes the first quantifier "there is" in the conclusion B and is therefore using the construction method to construct the value

A1: $y = \frac{4ac - b^2}{4a}$.

Interpretation of S2: *For a real number x, you have that*

$$ax^2 + bx + c = a\left(x + \frac{b}{2a}\right)^2 + \frac{4ac - b^2}{4a}.$$

The author is following the construction method and is showing that the value of y in $A1$ satisfies the something that happens associated with the quantifier "there is" in B, namely, that

B1: For all real numbers x, $ax^2 + bx + c \leq y$.

Recognizing the quantifier "for all" in $B1$, the author uses the choose method, as indicated by the words, "For a real number x ...." In other words, the author chooses

A2: a real number x,

for which it must be shown that

B2: $ax^2 + bx + c \leq y$.

In showing that $B2$ is true, the author first establishes the following statement in $S2$, omitting several algebraic steps:

A3: It follows that:

$$\begin{aligned} ax^2 + bx + c &= a\left(x^2 + \frac{b}{a}x + \frac{b^2}{4a^2}\right) + c - \frac{b^2}{4a} \\ &= a\left(x + \frac{b}{2a}\right)^2 + \frac{4ac - b^2}{4a}. \end{aligned}$$

Interpretation of S3: *Because $a < 0$, it follows that*

$$ax^2 + bx + c \leq \frac{4ac - b^2}{4a} = y.$$

The author is working forward from the first term on the right side of $A3$. Specifically, the author notes that $a\left(x + \frac{b}{2a}\right)^2 \leq 0$ because $a < 0$ by hypothesis and $\left(x + \frac{b}{2a}\right)^2 \geq 0$ and therefore

A4: $ax^2 + bx + c \leq \frac{4ac - b^2}{4a}$.

Finally, the author use the fact that $\frac{4ac - b^2}{4a} = y$, from $A1$. This completes both the choose and construction methods and so the proof is finished.

Summary

When working with statements containing nested quantifiers, follow these steps:

1. For each quantifier encountered from left to right, identify the object, the certain property, and the something that happens.

2. Apply appropriate techniques such as construction, choose, and specialization, based on the order of the quantifiers as they appear from left to right.

All material thus far is organized around the forward-backward method. Now it is time to see other techniques for showing that "A implies B."

Exercises

Note: Solutions to exercises marked with a B are in the back of this book. Solutions to exercises marked with a W are located on the World Wide Web at http://www.wiley.com/college/solow/.

Note: All proofs should contain an analysis of proof and a condensed version. Definitions for all mathematical terms are provided in the glossary at the end of the book.

^{B}**7.1**   Identify the objects, the certain property, and the something that happens for each of the quantifiers as they appear from left to right in each of the following definitions.

   a. The function f of one variable is **bounded above** if and only if there is a real number y such that for every real number x, $f(x) \leq y$.

   b. A set of real numbers S is **bounded** if and only if there is a real number $M > 0$ such that $\forall$ element $x \in S$, $|x| < M$.

   c. The function f of one variable is **continuous at the point** x if and only if for every real number $\epsilon > 0$, there is a real number $\delta > 0$ such that, for all real numbers y with $|x - y| < \delta$, $|f(x) - f(y)| < \epsilon$.

   d. Suppose that $x_1, x_2, \ldots$ are real numbers. The sequence $x_1, x_2, \ldots$ **converges to the real number** x if and only if $\forall$ real numbers $\epsilon > 0$, $\exists$ an integer $k' \ni \forall$ integer k with $k > k'$, $|x^k - x| < \epsilon$.

7.2   Rewrite each the following statements introducing appropriate notation and using nested quantifiers.

   a. A set S of real numbers has the property that, no matter which element is chosen in the set, you can find another element in the set that is strictly larger.

   b. A function of a single variable has the property that for some real number, the absolute value of the function is always less than that number.

^{B}**7.3**   Are each of the following pairs of statements the same, that is, are these two statements both true? Why or why not? Explain.

  a. $S1:$ For all real numbers x with $0 \leq x \leq 1$, for all real
        numbers y with $0 \leq y \leq 2$, $2x^2 + y^2 \leq 6$.
      $S2:$ For all real numbers y with $0 \leq y \leq 2$, for all real
        numbers x with $0 \leq x \leq 1$, $2x^2 + y^2 \leq 6$.

  b. $S1:$ For all real numbers x with $0 \leq x \leq 1$, for all real
        numbers y with $0 \leq y \leq 2x$, $2x^2 + y^2 \leq 6$.
      $S2:$ For all real numbers y with $0 \leq y \leq 1$, for all real
        numbers x with $0 \leq x \leq 2y$, $2x^2 + y^2 \leq 6$.

  c. Based on your answer to parts (a) and (b), when is the statement, "for all objects X with a certain property P and for all objects Y with a certain property Q, something happens," the same as the statement, "for all objects Y with the property Q and for all objects X with the property P, something happens"?

7.4   Are each of the following pairs of statements the same, that is, are these two statements both true? Why or why not? Explain.

  a. $S1:$ There is a real number $x \geq 2$ such that there is a
        real number $y \geq 1$ such that $x^2 + 2y^2 < 9$.
      $S2:$ There is a real number $y \geq 1$ such that there is a
        real number $x \geq 2$ such that $x^2 + 2y^2 < 9$.

  b. $S1:$ There is a real number $x \leq 1$ such that there is a
        real number $y \leq 2x$ such that $2x^2 + y^2 > 6$.
      $S2:$ There is a real number $y \leq 1$ such that there is a
        real number $x \leq 2y$ such that $2x^2 + y^2 > 6$.

  c. Based on your answers to parts (a) and (b), when is the statement, "there is an object X with a certain property P such that there is an object Y with a certain property Q such that something happens," the same as the statement, "there is an object Y with the property Q such that there is an object X with the property P such that something happens"?

^{B}**7.5**   For each of the following statements in the backward process, indicate which techniques you would use (construction and choose) and in which order. Also, explain how you would apply the techniques to the particular problem, that is, what would you construct, what would you choose, and so on.

  a. There is a real number $M > 0$ such that, for all elements x in the set T of real numbers, $|x| \leq M$.

b. For all real numbers $M > 0$, there is an element x in the set T of real numbers such that $|x| > M$.

c. $\forall$ real numbers $\epsilon > 0$, $\exists$ a real number $\delta > 0$ $\Rightarrow$ $\forall$ real numbers x and y with $|x - y| < \delta$, $|f(x) - f(y)| < \epsilon$ (where f is a function of one variable).

7.6 Explain how to work forward from each of the following statements. How would you apply the proof techniques?

a. For all objects X with a certain property P, there is an object Y with a certain property Q such that something happens.

b. There is an object X with a certain property P such that, for all objects Y with a certain property Q, something happens.

^{B}**7.7** Prove that for every real number $x > 2$, there is a real number $y < 0$ such that $x = 2y/(1 + y)$.

7.8 Prove that if $S = \{$real numbers $x > 0 : x^2 < 2\}$, then for every real number $\epsilon > 0$, there is an element $x \in S$ such that $x^2 > 2 - \epsilon$.

^{W}**7.9** Prove that the function $f(x) = -x^2 + 2x$ is bounded above, that is, there is a real number y such that for all real numbers x, $f(x) \leq y$.

7.10 Prove that if

$$\begin{aligned} S &= \{(x, y) : x^2 + y^2 \leq 1\} \quad \text{and} \\ T &= \{(x, y) : (x - 3)^2 + (y - 4)^2 \leq 1\}, \end{aligned}$$

then there are real numbers a and b such that for every $(x, y) \in S$, $y \leq ax + b$ and for every $(x, y) \in T$, $y \geq ax + b$. (Hint: Draw a picture of S and T and find an appropriate line $y = ax + b$.)

^{W}**7.11** Prove that the set $S = \{1 - 1/2, 1 - 1/3, 1 - 1/4, \ldots\}$ satisfies the property that for every $\epsilon > 0$, there is an $x \in S$ such that $x > 1 - \epsilon$. (Hint: You can write S as $\{$real numbers $x :$ there is an integer $n \geq 2$ such that $x = 1 - 1/n\}$.)

7.12 Let f and g be functions of one variable. Prove that if f and g are onto (see Definition 16), then the function $f \circ g$ is onto, where $(f \circ g)(x) = f(g(x))$.

$$8$$

The Contradiction Method

With all the techniques you have learned so far, you may well find yourself unable to complete a proof for one reason or another. This chapter presents a new technique that often provides a successful alternative. This method is used when the conclusion of the proposition contains appropriate key words.

8.1 WHY THE NEED FOR ANOTHER PROOF TECHNIQUE?

As powerful as the forward-backward method is, it may not always lead to a successful proof, as shown in the next example.

Proposition 12 *If n is an integer and n^2 is even, then n is even.*

Analysis of Proof. The forward-backward method gives rise to the key question, "How can I show that an integer (namely, n) is even?" One answer is to show that

B1: There is an integer k such that $n - 2k$.

The appearance of the quantifier "there is" in the backward process suggests proceeding with the construction method, and so the forward process is used in an attempt to produce the desired integer k.

Working forward from the hypothesis that n^2 is even, you can state that

A1: There is an integer, say, m, such that $n^2 = 2m$.

The objective is to produce an integer k for which $n = 2k$, so it is natural to take the positive square root of both sides of the equality in $A1$ to obtain

A2: $n = \sqrt{2m}$,

but how can you rewrite $\sqrt{2m}$ to look like $2k$? It would seem that the forward-backward method has failed.

Proof of Proposition 12.   The technique you are about to learn leads to a simple proof of this proposition, which is left as an exercise. $\square$

Fortunately, there are several other techniques that you might want to try before you give up. In this chapter, the **contradiction method** is described, together with an indication of how and when it should be used.

8.2   HOW AND WHEN TO USE THE CONTRADICTION METHOD

With the contradiction method, you begin by assuming that A is true, just as you do in the forward-backward method. However, to reach the desired conclusion that B is true, you proceed by asking yourself the simple question, "Why can't B be false?" After all, if B is supposed to be true, then there must be some reason why B cannot be false. The objective of the contradiction method is to discover that reason.

In other words, the idea of a proof by contradiction is to assume that A is true and B is false, and see why this cannot happen. So what does it mean to "see why this cannot happen"? Suppose, for example, that as a result of assuming that A is true and B is false (hereafter written as $NOT\ B$), you were somehow able to reach the conclusion that $1 = 0$!?! Would that not convince you that it is impossible for A to be true and B to be false simultaneously? Thus, in a proof by contradiction, you assume that A is true and that $NOT\ B$ is true. You must use this information to reach a contradiction to something that you absolutely know is true.

Another way of viewing the contradiction method is to recall from Table 1.1 on page 4 that the statement "A implies B" is true in all cases except when A is true and B is false. In a proof by contradiction, you rule out this one unfavorable case by assuming that it actually does happen, and then reaching a contradiction.

At this point, several natural questions arise:

1. What contradiction should you be looking for?

2. Exactly how do you use the assumption that A is true and B is false to reach the contradiction?

3. Why and when should you use this approach instead of the forward-backward method?

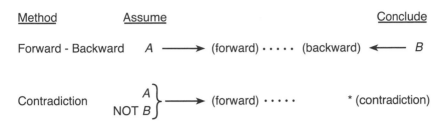

Fig. 8.1 Comparing the forward-backward and contradiction methods.

The first question is by far the hardest to answer because there are no specific guidelines. Each problem gives rise to its own contradiction. It takes creativity, insight, persistence, and sometimes luck to produce a contradiction.

As to the second question, the most common approach to finding a contradiction is to work forward from the assumptions that A and $NOT\ B$ are true, as is illustrated in a moment.

The foregoing discussion also indicates why you might wish to use contradiction instead of the forward-backward method. With the forward-backward method you assume only that A is true, while in the contradiction method, you assume that both A and $NOT\ B$ are true. Thus, you have two statements from which to reason forward instead of just one (see Figure 8.1). On the other hand, you do not know what the contradiction will be and therefore cannot work backward.

As a general rule, use contradiction when the statement $NOT\ B$ gives you useful information. There are at least two recognizable instances when this happens. Recall the statement B associated with Proposition 12, "n is an even integer." Because an integer is either odd or even, when you assume that B is not true (that is, that n is not an even integer), it must be the case that n is odd. Here, the statement $NOT\ B$ has given you some useful information. In general, when the statement B is one of two possible alternatives (as in Proposition 12), the contradiction method is likely to be effective because, by assuming $NOT\ B$, you will know that the other case must happen, and that should help you to reach a contradiction.

A second instance when the contradiction method is likely to be successful is when the statement B contains the key word "no" or "not," as shown now.

Proposition 13 *If r is a real number such that $r^2 = 2$, then r is irrational.*

Analysis of Proof. It is important to note that you can rewrite the conclusion of Proposition 13 to read "r is not rational." In this form, the appearance of the key word "not" now suggests using the contradiction method, whereby you assume that A and $NOT\ B$ are both true—in this case,

A: $r^2 = 2$,   and
A1 ($NOT\ B$)**:** r is a rational number.

A contradiction must now be reached using this information.

Working forward from $A1$ by using Definition 7 on page 24 for a rational number, you can state that

A2: There are integers p and q with $q \neq 0$ such that $r = p/q$.

There is still the unanswered question of where the contradiction arises, and this takes a lot of creativity. A crucial observation here really helps—it is possible to assume that

A3: p and q have no common divisor (that is, there is no integer that divides both p and q).

The reason for this is that if p and q did have a common divisor, you could divide this integer out of both p and q.

Now you can reach a contradiction by showing that 2 is a common divisor of p and q! This is done by working forward to show that p and q are even, and hence 2 divides them both.

Working forward by squaring both sides of the equality in $A2$, it follows that

A4: $r^2 = p^2/q^2$.

But from A you also know that $r^2 = 2$, so

A5: $2 = p^2/q^2$.

The rest of the forward process is mostly rewriting $A5$ via algebraic manipulations to reach the desired contradiction that both p and q are even integers. Those steps and their justifications are provided in the following table.

Statement	Reason
A6: $2q^2 = p^2$.	Multiply both sides of $A5$ by q^2.
A7: p^2 is even.	From $A6$ because p^2 is 2 times some integer, namely, q^2.
A8: p is even.	From Proposition 12.
A9: $p = 2k$, for some integer k.	Definition of an even integer.
A10: $2q^2 = (2k)^2 = 4k^2$.	Substitute $A9$ in $A6$.
A11: $q^2 = 2k^2$.	Divide $A10$ through by 2.
A12: q^2 is even.	From $A11$ because q^2 is 2 times some integer, namely, k^2.
A13: q is even.	From Proposition 12.

So, both p and q are even (see $A8$ and $A13$), and this contradicts $A3$, thus completing the proof.

Proof of Proposition 13.   Assume, to the contrary, that r is a rational number of the form p/q (where p and q are integers with $q \neq 0$) and that $r^2 = 2$. Furthermore it can be assumed that p and q have no common divisor for, if they did, this number could be canceled from both the numerator p and the denominator q. Because $r^2 = 2$ and $r = p/q$, it follows that $2 = p^2/q^2$, or equivalently, $2q^2 = p^2$. Noting that $2q^2$ is even, p^2, and hence p, are even. Thus, there is an integer k such that $p = 2k$. On substituting this value for p, one obtains $2q^2 = p^2 = (2k)^2 = 4k^2$ or, equivalently, $q^2 = 2k^2$. From this it then follows that q^2, and hence q, are even. Thus it has been shown that both p and q are even and have the common divisor 2. This contradiction establishes the claim. □

This proof, discovered in ancient times by a follower of Pythagoras, epitomizes the use of contradiction. Try to prove the statement by some other method.

8.3   ADDITIONAL USES FOR THE CONTRADICTION METHOD

There are several other valuable uses for the contradiction method. Recall that, when the statement B contains the quantifier "there is," the construction method is recommended in spite of the difficulty of actually having to produce the desired object. The contradiction method opens up a whole new approach. Instead of trying to show that there is an object with the certain property such that the something happens, why not proceed from the assumption that there is no such object? Now your job is to use this information to reach some kind of contradiction. How and where the contradiction arises is not at all clear, but finding a contradiction might be easier than producing or constructing the object. Consider the following example.

Suppose you wish to show that at a party of 367 people, there are at least two people whose birthdays fall on the same day of the year. If the construction method is used, then you would actually have to go to the party and find two such people. Using the contradiction method saves you the time and trouble of having to do so. With the contradiction method, you can assume that no two people's birthdays fall on the same day of the year, or equivalently, that everyone's birthday falls on a different day of the year.

To reach a contradiction, assign numbers to the people in such a way that the person with the earliest birthday of the year receives the number 1, the person with the next earliest birthday receives the number 2, and so on. Recall that each person's birthday is assumed to fall on a different day. Thus, the birthday of the person whose number is 2 must occur at least one day later than the person whose number is 1, and so on. Consequently, the birthday of the person whose number is 367 must occur at least 366 days after the person whose number is 1. But a year has at most 366 days, and so this is impossible—you have therefore reached a contradiction.

This example illustrates a subtle but significant difference between a proof using the construction method and one that uses contradiction. If the construction method is successful, then you have produced the desired object, or at least indicated how the object might be produced, perhaps with the aid of a computer. On the other hand, if you establish the same result by contradiction, then you know that the object exists, but have not physically constructed the object. For this reason, it is often the case that proofs done by contradiction are quite a bit shorter and easier than those done by construction because you do not have to create the desired object. You only have to show that the object's nonexistence is impossible. This difference has led to some great philosophical debates in mathematics. Moreover, an active area of current research consists of finding constructive proofs where previously only proofs by contradiction were known.

8.4   READING A PROOF

The process of reading and understanding a proof is demonstrated with the following proposition.

Proposition 14 *If m and n are odd integers, then the equation $x^2 + 2mx + 2n = 0$ has no rational root.*

Proof of Proposition 14.   (For reference purposes, each sentence of the proof is written on a separate line.)

S1:   Assume that the equation $x^2 + 2mx + 2n = 0$ has a rational root, say, $x = p/q$, where one of p and q is odd and $q \neq 0$.

S2:   It now follows that q is odd, for otherwise $p^2 = -2mqp - 2q^2n$, which would mean that p^2, and hence p, is even.

S3:   Note first that if m' and n' are odd, then $y^2 + 2m'y + 2n' = 0$ has no root that is odd, for if y is such a root, then y^2 is odd and also $y^2 = -2m'y - 2n'$ is even, which cannot happen.

S4:   Also, $y^2 + 2m'y + 2n' = 0$ has no root that is even, for if y is such a root, then $y = 2k$, for some integer k.

S5:   But then $4k^2 + 4m'k + 2n' = 0$, that is, $2k^2 + 2m'k + n' = 0$, which cannot happen because $2k^2 + 2m'k$ is even and n' is odd.

S6:   However, the fact that $x = p/q$ satisfies $x^2 + 2mx + 2n = 0$ means that p satisfies $y^2 + 2m'y + 2n'$, where $m' = mq$ is odd and $n' = nq^2$ is also odd, which cannot happen.

This contradiction establishes the claim. $\Box$

Analysis of Proof. An interpretation of statements $S1$ through $S6$ follows.

Interpretation of S1: *Assume that the equation $x^2 + 2mx + 2n = 0$ has a rational root, say, $x = p/q$, where one of p and q is odd and $q \neq 0$.*
    The author is assuming that the conclusion is not true, that is:

A1 (*NOT B*): There is a rational root $x = p/q$ to the equation,

thus indicating that this is a proof by contradiction. Note that the author has assumed further that

A2: One of p and q is odd.

This is justified because if both are even, then you can repeatedly divide both p and q by 2 until one of them is odd.
    Having recognized that this is a proof by contradiction, did you read ahead to identify the contradiction? If not, then you should do so now.

Interpretation of S2: *It now follows that q is odd, for otherwise $p^2 = -2mqp - 2q^2n$, which would mean that p^2, and hence p, is even.*
    The author is working forward from $A1$ to claim that

A3: q is odd.

The author reaches this conclusion by contradiction. That is, by assuming that q is even, the author claims that p is also even, which contradicts $A2$. Indeed this is correct because the author has substituted $x = p/q$ into $x^2 + 2mx + 2n = 0$, multiplied through by q^2, and solved for p^2 to obtain $p^2 = -2mqp - 2q^2n$. Now you can see that the right side is even and so p^2 is even. Finally, Proposition 12 ensures that p is also even.

Interpretation of S3: *Note first that if m' and n' are odd, then $y^2 + 2m'y + 2n' = 0$ has no root that is odd, for if y is such a root, then y^2 is odd and also $y^2 = -2m'y - 2n'$ is even, which cannot happen.*
    This statement is directed toward reaching the final contradiction. The author is establishing that

A4: If m' and n' are odd integers, then $y^2 + 2m'y + 2n' = 0$
    has no root that is an odd integer.

This also is shown by contradiction, which is indicated when the author assumes that y is an odd integer root. The author then reaches the contradiction that y^2 is both odd (because odd times odd is odd) and even (because, from $A4$, $y^2 = -2m'y - 2n'$, which is even).

Interpretation of S4: *Also, $y^2 + 2m'y + 2n' = 0$ has no root that is even, for if y is such a root, then $y = 2k$, for some integer k.*
    Here, the author is establishing that

> **A5:** If m' and n' are odd, then $y^2 + 2m'y + 2n' = 0$ has no root that is even.

This is also shown by contradiction, which is indicated when the author assumes that y is an even integer root. By definition, this means that $y = 2k$, for some integer k.

Interpretation of S5: *But then $4k^2 + 4m'k + 2n' = 0$, that is, $2k^2 + 2m'k + n' = 0$, which cannot happen because $2k^2 + 2m'k$ is even and n' is odd.*
The author is continuing the contradiction argument started in *S4*. In particular, the author substitutes $y = 2k$ into $y^2 + 2m'y + 2n' = 0$ and divides the result by 2 to obtain $2k^2 + 2m'k + n' = 0$. The author then claims correctly that a contradiction is reached because it is impossible for the even integer $2k^2 + 2m'k$ plus the odd integer n' to be 0.

Interpretation of S6: *However, the fact that $x = p/q$ satisfies $x^2 + 2mx + 2n = 0$ means that p satisfies $y^2 + 2m'y + 2n'$, where $m' = mq$ is odd and $n' = nq^2$ is also odd, which cannot happen.*
The author is working forward from $A1$ by substituting $x = p/q$ into $x^2 + 2mx + 2n = 0$, multiplying through by q^2, and then realizing that

> **A6:** p satisfies the equation $y^2 + 2m'y + 2n'$, where $m' = mq$ and $n' = nq^2$.

The author further claims correctly that both $m' = mq$ and $n' = nq^2$ are odd [because m and n are odd (see the hypothesis), and q is also odd (see $A3$)].
Finally, the author claims to have reached the desired contradiction. Indeed, $A4$ and $A5$ together mean that the equation $y^2 + 2m'y + 2n' = 0$, where m' and n' are odd, has no integer solution. Yet in $A6$, the author shows that p is an integer solution to such an equation. Thus the proof is complete.

Summary

The contradiction method is a useful technique when the statement B contains the key word "no" or "not." To use this method, follow these steps:

1. Assume that A is true and B is not true (that is, assume that A and *NOT B* are true).

2. Work forward from A and *NOT B* to reach a contradiction.

One of the disadvantages of this method is that you do not know exactly what the contradiction is going to be. The next chapter describes another proof technique in which you attempt to reach a specific contradiction. As such, you will have a "guiding light" because you will know what contradiction you are looking for.

Exercises

Note: Solutions to exercises marked with a B are in the back of this book. Solutions to exercises marked with a W are located on the World Wide Web at http://www.wiley.com/college/solow/.

Note: All proofs should contain an analysis of proof and a condensed version. Definitions for all mathematical terms are provided in the glossary at the end of the book.

^{B}**8.1**   When applying the contradiction method to the following propositions, what should you assume?

    a. If l, m, and n are three consecutive integers, then 24 does not divide $l^2 + m^2 + n^2 + 1$.

    b. If the matrix M is not singular, then the rows of M are not linearly dependent.

    c. If f and g are two functions such that (1) $g \geq f$ and (2) f is unbounded above, then g is unbounded above.

8.2   Consider applying contradiction to show that if a and b are integers and b is odd, then ± 1 are not roots of $ax^2 + bx + a$.

    a. What statement(s) would you work backward from?

    b. At the end of the proof, a mathematics student said, "... and because I have been able to show that b is even, the proof is complete." Do you agree with the student? Why or why not? Explain.

^{B}**8.3**   Reword each of the following statements so that the word "not" appears explicitly.

    a. There are an infinite number of primes.

    b. The only positive integers that divide the positive integer p are 1 and p.

    c. The lines l and l' in a plane P are parallel.

8.4   Reword each of the following statements so that the word "not" does not appear.

    a. The real number $ad - bc$ is not equal to 0.

    b. The triangle ABC is not equilateral.

    c. The polynomial $a_0 + a_1 x + \cdots + a_n x^n$ has no real root.

^{B}**8.5**   When trying to prove each of the following statements, which techniques would you use and in which order? Specifically, state what you would assume and what you would try to conclude. (Throughout, S and T are sets of real numbers and all of the variables refer to real numbers.)

   a. $\exists s \in S \ni s \in T$.

   b. $\forall s \in S$, $\nexists t \in T$ such that $s > t$.

   c. $\nexists M > 0$ such that, $\forall x \in S$, $|x| < M$.

8.6   When trying to prove each of the following statements, which techniques would you use and in which order? Specifically, state what you would assume and what you would try to conclude. (Throughout, S is a given set of real numbers and all of the variables refer to real numbers.)

   a. For every real number $\epsilon > 0$, there is an element $x \in S$ such that $x > u - \epsilon$ (where u is a given real number).

   b. There is a real number $y > 0$ such that for every element $x \in S$, $f(x) < y$ (where f is a function of one variable).

   c. For every line l in the plane that is parallel to, but different from, l', there is no point on l that is also in S (where l' is a given line in the plane).

^{B}**8.7**   Prove, by contradiction, that if n is an integer and n^2 is even, then n is even.

8.8   Prove, by contradiction, that there do not exist positive real numbers x and y with $x \neq y$ such that $x^3 - y^3 = 0$.

^{B}**8.9**   Prove, by contradiction, that no chord of a circle is longer than a diameter.

8.10   Suppose that a, b, and c are real numbers with $c \neq 0$. Prove, by contradiction, that if $cx^2 + bx + a$ has no rational root, then $ax^2 + bx + c$ has no rational root.

^{W}**8.11**   Prove, by contradiction, that if $n - 1$, n, and $n + 1$ are consecutive positive integers, then the cube of the largest cannot be equal to the sum of the cubes of the other two.

8.12   Prove that if p and q are integers with $p \neq q$ and p is prime and divides q, then q is not prime.

^{B}**8.13**   Prove, by contradiction, that at a party of $n \geq 2$ people, there are at least two people who have the same number of friends at the party.

8.14   Prove, by contradiction, that there are at least two people on the planet who were born on the same second of the same hour of the same day of the same year in the twentieth century. (You can assume that there were at least 4 billion people born in that century.)

^{W}**8.15**   Prove, by contradiction, that there are an infinite number of primes. (Hint: Assume that n is the largest prime. Then consider any prime number p that divides $n! + 1$. How is p related to n?)

^{B}**8.16**   Identify the contradiction in the following proof.

> **Proposition.** If a and b are integers and b is odd, then ± 1 are not roots of $ax^4 + bx^2 + a$.
>
> **Proof.**   Assume, to the contrary, that $+1$ or -1 is a root of $ax^4 + bx^2 + a$. Then $a(\pm 1)^4 + b(\pm 1)^2 + a = 0$, that is, $b + 2a = 0$. Thus, $b = -2a$, which cannot happen, and so the proof is complete.   □

8.17   Identify the contradiction in the following proof.

> **Proposition.** If a and b are integers with $a \neq 0$ and the number of rational roots of $ax^4 + bx^2 + a$ is odd, then b is even.
>
> **Proof.**   Assume, to the contrary, that b is odd. Then, by the proposition in the previous exercise, ± 1 are not roots of $ax^4 + bx^2 + a$. Now consider a rational root p/q of $ax^4 + bx^2 + a$. Note that $p \neq 0$ for otherwise $a = 0$. It is easy to verify that q/p is also a rational root of $ax^4 + bx^2 + a$. Thus, rational roots come in pairs. Because the number of rational roots of $ax^4 + bx^2 + a$ is odd, it must be that one of these roots is repeated, so, $p/q = q/p$. But then $p = \pm q$, which cannot happen, and so the proof is complete.   □

^{B}**8.18**   Write an analysis of proof that corresponds to the condensed proof given below. Indicate which techniques are used and how they are applied. Fill in the details of any missing steps where appropriate.

> **Proposition.** If a, b, c, x, y, and z are real numbers with $b \neq 0$ such that (1) $az - 2by + cx = 0$ and (2) $ac - b^2 > 0$, then it must be that $xz - y^2 \leq 0$.
>
> **Proof.** Assume that $xz - y^2 > 0$. From this and (2) it follows that $(ac)(xz) > b^2 y^2$. Rewriting (1) and squaring both sides, one obtains $(az + cx)^2 = 4b^2 y^2 < 4(ac)(xz)$. Rewriting, one has that $(az - cx)^2 < 0$, which cannot happen.   □

^{W}**8.19**   Write an analysis of proof that corresponds to the condensed proof given below. Indicate which techniques are used and how they are applied. Fill in the details of any missing steps where appropriate.

Proposition. The polynomial $x^4 + 2x^2 + 2x + 2$ cannot be expressed as the product of the two polynomials $x^2 + ax + b$ and $x^2 + cx + d$ in which a, b, c, and d are integers.

Proof. Suppose that

$$x^4 + 2x^2 + 2x + 2 = (x^2 + ax + b)(x^2 + cx + d)$$

$$= x^4 + (a + c)x^3 + (b + ac + d)x^2 + (bc + ad)x + bd.$$

It would then follow that the integers a, b, c, and d satisfy

1. $a + c = 0$.
2. $b + ac + d = 2$.
3. $bc + ad = 2$.
4. $bd = 2$.

The only way (4) can happen is if one of the factors b or d is odd (± 1) and the other is even (± 2). Suppose that b is in fact odd and d is even. From (3), it would then follow that c is even, but then the left side of (2) would be odd, which is impossible. A similar contradiction can be reached if b is even and d is odd. $\square$

9

The Contrapositive
Method

In the contradiction method described in the previous chapter, you work forward from the two statements A and NOT B to reach a contradiction. The challenge with this method is to determine what the contradiction is going to be. The technique presented in this chapter has the advantage of directing you toward one specific contradiction.

9.1 HOW AND WHEN TO USE THE CONTRAPOSITIVE METHOD

The **contrapositive method** is similar to contradiction in that you begin by assuming that A and NOT B are true. Unlike contradiction, however, you do not work forward from both A and NOT B. Instead, you work forward only from NOT B. Your objective is to reach the contradiction that A is false (hereafter written NOT A). Can you ask for a better contradiction than that? How can A be true and false at the same time? To repeat, in the contrapositive method you assume that A and NOT B are true; you then work forward from NOT B to reach the contradiction that A is false.

You can think of the contrapositive method as a passive form of contradiction in the sense that the assumption that A is true passively provides the contradiction. In the contradiction method, the assumption that A is true is used actively to reach a contradiction. The next proposition demonstrates the contrapositive method.

Definition 17 *A function f of one variable is **one-to-one** if and only if for all real numbers x and y with $x \neq y$, $f(x) \neq f(y)$.*

Proposition 15 *If m and b are real numbers with $m \neq 0$, then the function $f(x) = mx + b$ is one-to-one.*

Analysis of Proof.   The forward-backward method is used to begin the proof because the proposition does not contain any key words (such as "for all," "there is," "no," "not," and so on). The key question associated with B is, "How can I show that a function is one-to-one?" Applying Definition 17 to the particular function $f(x) = mx + b$ means you must show that

B1:   For all real numbers x and y with $x \neq y$, $mx + b \neq my + b$.

This new statement contains both the key words "for all" and "not" (in "not equal"). When more than one group of key words is present in a statement, apply appropriate techniques based on the occurrence of the key words as they appear from left to right (just as you learned to do with nested quantifiers in Chapter 7). Because the first key words from the left in the backward statement $B1$ are "for all," the choose method is used next. So, choose

A1:   Real numbers x and y with $x \neq y$,

for which you must show that

B2: $mx + b \neq my + b$.

Recognizing the key word "not" in $B2$, it is appropriate to proceed with either the contradiction or contrapositive method. In this case, the contrapositive method is used. Thus, you work forward from *NOT B2*:

A2 (*NOT B2*): $mx + b = my + b$.

and backward from *NOT A1* (the last statement in the forward process):

B3 (*NOT A1*): $x = y$.

The remainder of the proof is simple algebra applied to $A2$. Specifically, on subtracting b from both sides of $A2$ you obtain:

A3: $mx = my$.

Finally, because $m \neq 0$ by the hypothesis A, you can divide both sides of $A3$ by m and obtain $B3$, and so the proof is complete.

In the condensed proof that follows, note that no mention is made of the choose or contrapositive methods.

Proof of Proposition 15.   Let x and y be real numbers for which $mx + b = my + b$. It is shown that $x = y$. But this follows by subtracting b from both sides and then dividing by m, noting that $m \neq 0$. $\square$

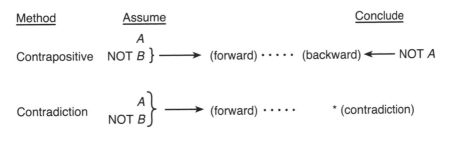

Fig. 9.1 Comparing the contrapositive and contradiction methods.

9.2 COMPARING THE CONTRAPOSITIVE METHOD

As already mentioned, the contrapositive method is a type of proof by contradiction. Each of these methods has its advantages and disadvantages, as illustrated in Figure 9.1. The disadvantage of the contrapositive method is that you work forward from only one statement (namely, *NOT B*) instead of two. On the other hand, the advantage is that you know what you are looking for (namely, *NOT A*). Thus you can apply the backward process to *NOT A*. The option of working backward is not available in the contradiction method because you do not know what contradiction you are looking for.

It is also interesting to compare the contrapositive and forward-backward methods. In the forward-backward method, you work forward from *A* and backward from *B*; in the contrapositive method, you work forward from *NOT B* and backward from *NOT A* (see Figure 9.2).

From Figure 9.2, it is not hard to see why the contrapositive method might be better than the forward-backward method. Perhaps you can obtain more useful information by working forward from *NOT B* rather than from *A*. It might also be easier to work backward from *NOT A* rather than from *B*, as is done in the forward-backward method.

The forward-backward method arises from considering what happens to the truth of "*A* implies *B*" when *A* is true and when *A* is false (recall Table 1.1 on page 4). The contrapositive method arises from similar considerations regarding *B*. Specifically, if *B* is true then, according to Table 1.1, the statement "*A* implies *B*" is true. Hence, there is no need to consider the case when *B* is

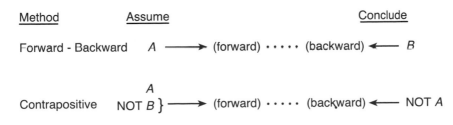

Fig. 9.2 Comparing the forward-backward and contrapositive methods.

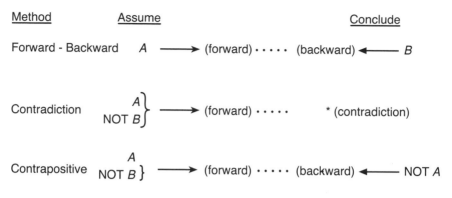

Fig. 9.3 Comparing the forward-backward, contradiction, and contrapositive methods.

true. So suppose B is false. To ensure that "A implies B" is true, according to Table 1.1, you must show that A is false. Thus, with the contrapositive method, you assume that B is false and try to conclude that A is false.

The statement "A implies B" is logically equivalent to "$NOT\ B$ implies $NOT\ A$" (see Table 3.2 on page 30). You can therefore think of the contrapositive method as the forward-backward method applied to the statement "$NOT\ B$ implies $NOT\ A$." Most condensed proofs that use the contrapositive method make no reference to the contradiction method.

There is one instance that indicates you should use, or at least consider seriously, the contradiction or contrapositive method. This occurs when B contains the key word "no" or "not," for then, $NOT\ B$ has some useful information.

In the contradiction method, you work forward from the two statements A and $NOT\ B$ to obtain a contradiction. In the contrapositive method, you also reach a contradiction, but you do so by working forward from $NOT\ B$ to reach the conclusion $NOT\ A$; you should also work backward from $NOT\ A$. A comparison of the forward-backward, contradiction, and contrapositive methods is given in Figure 9.3.

9.3 READING A PROOF

The process of reading and understanding a proof is demonstrated with the following proposition.

Proposition 16 *Assume that a and b are integers with $a \neq 0$. If a does not divide b, then $ax^2 + bx + b - a$ has no positive integer root.*

Proof of Proposition 16.   (For reference purposes, each sentence of the proof is written on a separate line.)

S1:   Suppose that $x > 0$ is an integer with $ax^2 + bx + b - a = 0$.
S2:   Then $x = \frac{-b \pm (b - 2a)}{2a}$.
S3:   Because $x > 0$, it must be that $x = 1 - \frac{b}{a}$.
S4:   But then $b = (1 - x)a$ and so $a|b$.

The proof is now complete.   □

Analysis of Proof.   An interpretation of statements $S1$ through $S4$ follows.

Interpretation of S1: *Suppose that $x > 0$ is an integer with*
$ax^2 + bx + b - a = 0$.
    The author is assuming $NOT\ B$, that is,

A1 ($NOT\ B$): $x > 0$ is an integer with $ax^2 + bx + b - a = 0$.

This means that the contradiction or contrapositive method is used. However, on reading $S4$, you see that the author reaches the statement $NOT\ A$:

B1 ($NOT\ A$): $a|b$,

thus indicating that this is a proof by the contrapositive method.

Interpretation of S2: *Then $x = \frac{-b \pm (b - 2a)}{2a}$.*
    The author works forward from $A1$ using the quadratic formula to obtain the following:

A2: $x = \frac{-b \pm \sqrt{b^2 - 4a(b-a)}}{2a} = \frac{-b \pm (b - 2a)}{2a}$.

Interpretation of S3: *Because $x > 0$, it must be that $x = 1 - \frac{b}{a}$.*
    The author works forward by algebra from $A2$ to see that the two roots are $x = -1$ and $x = 1 - \frac{b}{a}$. Because, from $A1$, $x > 0$, the author rules out the root $x = -1$ and correctly concludes that

A3: $x = 1 - \frac{b}{a}$.

Interpretation of S4: *But then $b = (1 - x)a$ and so $a|b$.*
    The author works backward from $B1$ by asking the key question, "How can I show that an integer (namely, a) divides another integer (namely, b)?" Using the definition to answer the question, the author needs to show the following:

B2: There is an integer c such that $b = ca$.

Recognizing the quantifier "there is" in $B2$, the author uses the construction method to produce the value of c. Specifically, the author works forward from $A3$ by algebra to solve for b, obtaining $b = (1 - x)a$, from which the author notes—without mentioning—that the desired value of c in $B2$ is $c = 1 - x$. Having reached $NOT\ A$, the contrapositive method is now complete and so, therefore, is the proof.

Summary

The contrapositive method, being a type of proof by contradiction, is used when the last statement in the backward process contains the key word "no" or "not." With the contrapositive method, you work toward a specific, known contradiction by

1. Assuming that A and NOT B are true.

2. Working forward from NOT B in an attempt to obtain NOT A.

3. Working backward from NOT A in an attempt to reach NOT B.

You can also think of the contrapositive method as the forward-backward method applied to the statement "NOT B implies NOT A" because this statement is logically equivalent to the statement "A implies B."

Both the contrapositive and contradiction methods require that you be able to write the NOT of a statement. In the next chapter you will learn how to do so when the statements contain key words, such as the quantifiers.

Exercises

Note: Solutions to exercises marked with a B are in the back of this book. Solutions to exercises marked with a W are located on the World Wide Web at http://www.wiley.com/college/solow/.

Note: All proofs should contain an analysis of proof and a condensed version. Definitions for all mathematical terms are provided in the glossary at the end of the book.

^{B}**9.1** If the contrapositive method is used to prove the following propositions, then what statement(s) will you work forward from and what statement(s) will you work backward from?

a. If n is an integer for which n^2 is even, then n is even.

b. Suppose that S is a subset of the set T of real numbers. If S is not bounded, then T is not bounded.

c. If the matrix M is not singular, then the rows of M are not linearly dependent.

9.2 Bob said, "If I study hard enough, then I will get at least a B in this course," to which Mary said "You got an A in the course, so therefore you studied hard enough." Was Mary's statement correct? Why or why not? Explain.

^{B}**9.3** In a proof by the contrapositive method of the proposition, "If r is a real number with $r > 1$, then there is no real number t with $0 < t < \pi/4$

such that $\sin(t) = r\cos(t)$," which of the following is a result of the forward process?

a. $r - 1 \geq 0$.

b. $\sin^2(t) = r^2(1 - \sin^2(t))$.

c. $1 - r < 0$.

d. $\tan(t) = 1/r$.

9.4   Suppose that a, b, and c are real numbers with $a > 0$ and that you want to prove by the contrapositive method that, "If there is no real number x with $ax^2 + bx + c = 0$, then it is not true that for all real numbers x, $ax^2 + bx + c < 0$." Which of the construction, choose, and specialization techniques would you use subsequently in doing the proof? Explain.

^{B}**9.5**   Suppose the contrapositive method is used to prove the proposition, "If the derivative of the function f at the point x is not equal to 0, then x is not a local maximum of f." Which of the following is the correct key question? What is wrong with the other choices?

a. How can I show that the point x is a local maximum of the function f?

b. How can I show that the derivative of the function f at the point x is 0?

c. How can I show that a point is a local maximum of a function?

d. How can I show that the derivative of a function at a point is 0?

9.6   Suppose that f is a function of one variable, S is a set of real numbers, and that the contrapositive method is used to prove the proposition, "If no element $x \in S$ satisfies the property that $f(x) = 0$, then f is not bounded above." Which of the following is a correct key question? What is wrong with the other choices?

a. How can I show that a function is not bounded above?

b. How can I show that there is an element $x \in S$ such that $f(x) = 0$?

c. How can I show that a function is bounded above?

d. How can I show that there is a point in a set where the value of a function is 0?

^{B}**9.7**   Suppose that p and q are positive real numbers. Prove, by the contrapositive method, that if $\sqrt{pq} \neq (p + q)/2$, then $p \neq q$.

9.8   Suppose that a and b are positive real number. Prove, by the contrapositive method, that if $a \neq b$, then $(a + b)/2 > \sqrt{ab}$.

[W] **9.9**   Prove, by the contrapositive method, that if c is an odd integer, then the equation $n^2 + n - c = 0$ has no integer solution for n.

9.10   Use the approach in the proof of Proposition 15 to prove that the function $f(x) = x^3$ is one-to-one.

[W] **9.11**   Prove, by the contrapositive method, that if no angle of a quadrilateral $RSTU$ is obtuse, then the quadrilateral $RSTU$ is a rectangle.

9.12   Suppose that v is an upper bound for a set S of real numbers. Prove, by the contrapositive method, that if it is not true that there is a real number $\epsilon > 0$ such that for every element $x \in S$, $x \leq v - \epsilon$, then there is no real number $u < v$ such that u is an upper bound for S.

[B] **9.13**   Write an analysis of proof that corresponds to the condensed proof given below. Indicate which techniques are used and how they are applied. Fill in the details of any missing steps where appropriate.

> **Proposition.** Suppose that p is a positive integer. If there is no integer m with $1 < m \leq \sqrt{p}$ such that $m|p$, then p is prime.
>
> **Proof.**   Assume that p is not prime, that is, that there is an integer n with $1 < n < p$ such that $n|p$. In the event that $n \leq \sqrt{p}$, then n is the desired integer m. Otherwise, $n > \sqrt{p}$. Because $n|p$, there is an integer k such that $p = nk$. It then follows that k is the desired value for m. This is because $k > 1$, for otherwise, $p \leq n$. Also, $k \leq \sqrt{p}$, for otherwise, $nk > \sqrt{p}\sqrt{p} = p$. $\square$

9.14   Write an analysis of proof that corresponds to the condensed proof given below. Indicate which techniques are used and how they are applied. Fill in the details of any missing steps where appropriate.

> **Definition.** A set S of real numbers is **bounded** if and only if there is a real number $M > 0$ such that, for all elements $x \in S$, $|x| < M$.
>
> **Proposition.** Suppose that S and T are sets of real numbers with $S \subseteq T$. If S is not bounded, then T is not bounded.
>
> **Proof.**   Suppose that T is bounded. Hence, there is a real number $M' > 0$ such that, for all $x \in T$, $|x| < M'$. It is shown that S is bounded. To that end, let $x' \in S$. Because $S \subseteq T$, it follows that $x' \in T$. But then $|x'| < M'$ and so S is bounded, thus completing the proof. $\square$

10

Nots of Nots
Lead to Knots

As you saw in the previous chapter, the contrapositive method is a valuable proof technique. To use that method, however, you must be able to write $NOT\ B$ so that you can work forward from that statement. Similarly, you have to know exactly what $NOT\ A$ is so that you can work backward from that statement. This chapter provides rules for writing the NOT of various statements that contain key words, such as quantifiers.

10.1 WRITING THE NOT OF STATEMENTS HAVING NOT, AND, AND OR

In some instances, the NOT of a statement is easy to find. For example, if A is the statement, "the real number $x > 0$," then the NOT of A is, "it is not the case that the real number $x > 0$," or equivalently, "the real number x is not > 0." You can eliminate the word "not" altogether by incorporating this word into the statement to obtain "the real number $x \leq 0$."

When taking the NOT of a statement that already contains the word "no" or "not," the result is that the NOT cancels the existing "not." For example, if A is the statement,

    A: There is no integer x such that $x^2 + x - 11 = 0$,

then the NOT of this statement is,

    NOT A: There is an integer x such that $x^2 + x - 11 = 0$.

Special rules apply when writing the *NOT* of a statement that contains the words *AND* or *OR*. Specifically,

$$NOT \ [A \ AND \ B] \quad \text{becomes} \quad [NOT \ A] \ OR \ [NOT \ B],$$
$$NOT \ [A \ OR \ B] \quad \text{becomes} \quad [NOT \ A] \ AND \ [NOT \ B].$$

For example,

$$NOT \ [(x \geq 3) \ AND \ (y < 2)] \quad \text{becomes} \quad [(x < 3) \ OR \ (y \geq 2)],$$
$$NOT \ [(x \geq 3) \ OR \ (y < 2)] \quad \text{becomes} \quad [(x < 3) \ AND \ (y \geq 2)].$$

10.2   WRITING THE *NOT* OF A STATEMENT WITH QUANTIFIERS

A more challenging situation arises when the statements contain quantifiers. For instance, suppose that the statement B contains the quantifier "for all" in the standard form:

B: For all "objects" with a "certain property," "something happens."

Then the *NOT* of this statement is:

NOT B: It is not the case that, for all "objects" with the "certain property," "something happens,"

which really means that

NOT B: There is an object with the certain property for which the something does not happen.

Similarly, if the statement B contains the quantifier "there is" in the standard form:

B: There is an "object" with the "certain property" such that "something happens,"

then the *NOT* of this statement is:

NOT B: It is not the case that there is an "object" with the "certain property" such that "something happens,"

or, in other words,

NOT B: For all objects with the certain property, the something does not happen.

Notice that when a statement contains a quantifier, the *NOT* of that statement contains the "opposite" quantifier, that is, "for all" becomes "there is" and vice versa. In general, there are three steps to finding the *NOT* of a statement containing one or more quantifiers:

Steps for Finding the *NOT* of a Statement with Quantifiers

Step 1. Put the word *NOT* in front of the entire statement.

Step 2. If the word *NOT* appears to the left of a quantifier, then move the word *NOT* to the right of the quantifier and place the *NOT* just before the something that happens. As you do so, change the quantifier to its opposite—"for all" becomes "there is" and "there is" becomes "for all."

Step 3. When all of the quantifiers appear to the left of the *NOT*, eliminate the *NOT* by incorporating the *NOT* into the statement that appears immediately to its right.

These steps are demonstrated with the following examples.

1. For every real number $x \geq 2$, $x^2 + x - 6 \geq 0$.
   Step 1. *NOT* [for every real number $x \geq 2$, $x^2 + x - 6 \geq 0$]
   Step 2. There is a real number $x \geq 2$ such that
            NOT $[x^2 + x - 6 \geq 0]$.
   Step 3. There is a real number $x \geq 2$ such that $x^2 + x - 6 < 0$.

Note in Step 2 that, when the *NOT* is passed from left to right, the quantifier changes but the certain property (namely, $x \geq 2$) does not. Also, because the quantifier "for every" is changed to "there exists," it becomes necessary to replace the comma by the words "such that." If the quantifier "there exists" is changed to "for all," then the words "such that" are replaced with a comma, as is illustrated in the next example.

2. There is a real number $x \geq 2$ such that $x^2 + x - 6 \geq 0$.
   Step 1. *NOT* [there is a real number $x \geq 2$ such that
            $x^2 + x - 6 \geq 0$.]
   Step 2. For all real numbers $x \geq 2$, *NOT* $[x^2 + x - 6 \geq 0]$.
   Step 3. For all real numbers $x \geq 2$, $x^2 + x - 6 < 0$.

If the statement you are taking the *NOT* of contains nested quantifiers (see Chapter 7), then Step 2 is performed on each quantifier, in turn, as it appears from left to right. Step 2 is repeated until all quantifiers appear to the left of the *NOT*, as demonstrated in the next two examples.

3. For every real number x between -1 and 1, there is a real number y between -1 and 1 such that $x^2 + y^2 \leq 1$.

Step 1.   *NOT* [for every real number x between -1 and 1, there is a real number y between -1 and 1 such that $x^2 + y^2 \leq 1$.]

Step 2.   There is a real number x between -1 and 1 such that *NOT* [there is a real number y between -1 and 1 such that $x^2 + y^2 \leq 1$].

Step 2.   There is a real number x between -1 and 1 such that, for all real numbers y between -1 and 1, *NOT* [$x^2 + y^2 \leq 1$].

Step 3.   There is a real number x between -1 and 1 such that, for all real numbers y between -1 and 1, $x^2 + y^2 > 1$.

4. There is a real number x between -1 and 1 such that, for all real numbers y between -1 and 1, $x^2 + y^2 \leq 1$.

Step 1.   *NOT* [there is a real number x between -1 and 1 such that, for all real numbers y between -1 and 1, $x^2 + y^2 \leq 1$.]

Step 2.   For all real numbers x between -1 and 1, *NOT* [for all real numbers y between -1 and 1, $x^2 + y^2 \leq 1$].

Step 2.   For all real numbers x between -1 and 1, there is a real number y between -1 and 1 such that *NOT* [$x^2 + y^2 \leq 1$].

Step 3.   For all real numbers x between -1 and 1, there is a real number y between -1 and 1 such that $x^2 + y^2 > 1$.

Summary

The following list summarizes the rules for taking the *NOT* of statements that have a special form.

1. *NOT* [*NOT A*]   becomes   *A*.

2. *NOT* [*A AND B*]   becomes   [(*NOT A*) *OR* (*NOT B*)].

3. *NOT* [*A OR B*]   becomes   [(*NOT A*) *AND* (*NOT B*)].

4. *NOT* [there is an object with a certain property such that something happens] becomes "For all objects with the certain property, the something does not happen."

5. *NOT* [for all objects with a certain property, something happens] becomes "There is an object with the certain property such that the something does not happen."

In the event that a statement contains nested quantifiers, the word *NOT* is processed through each quantifier, from left to right.

Exercises

Note: Solutions to exercises marked with a B are in the back of this book. Solutions to exercises marked with a W are located on the World Wide Web at http://www.wiley.com/college/solow/.

Note: All proofs should contain an analysis of proof and a condensed version. Definitions for all mathematical terms are provided in the glossary at the end of the book.

^{B}**10.1**   Write the *NOT* of each of the definitions in Exercise 5.1 on page 54, that is, take the *NOT* of the statement that constitutes the definition of the term being defined. For instance, the *NOT* of the definition in Exercise 5.1(a) is, "the real number x^* is not a maximum of the function f if there is a real number x such that $f(x) > f(x^*)$."

10.2   Write the *NOT* of each of the following definitions in such a way that the word "not" does not appear.

a. A sequence $x_1, x_2, \ldots$ of real numbers is **increasing** if and only if for every integer $k = 1, 2, \ldots,$ $x_k < x_{k+1}$.

b. A sequence $x_1, x_2, \ldots$ of real numbers is **decreasing** if and only if for every integer $k = 1, 2, \ldots,$ $x_k > x_{k+1}$.

c. A sequence $x_1, x_2, \ldots$ of real numbers is **strictly monotone** if and only if the sequence is increasing or decreasing.

d. An integer d is the **greatest common divisor** of the integers a and b if and only if (i) $d|a$ and $d|b$ and (ii) whenever c is an integer for which $c|a$ and $c|b$, it follows that $c|d$.

e. The real number $f'(\bar{x})$ is the **derivative** of the function f at the point $\bar{x}$ if and only if $\forall$ real number $\epsilon > 0$, $\exists$ a real number $\delta > 0$ such that $\forall$ real number x with $0 < |x - \bar{x}| < \delta$, it follows that $|f(x) - f(\bar{x})|/|x - \bar{x}| < \epsilon$.

^{B}**10.3**   Reword the following statements so that the word "not" appears explicitly. For example, reword the statement "$x > 0$" to read "x is not ≤ 0."

a. For each element x in the set S, x is in T.

b. There is an angle t between 0 and $\pi/2$ such that $\sin(t) = \cos(t)$.

c. For every object with a certain property, something happens.

d. There is an object with a certain property such that something happens.

10.4   What would you assume if the contradiction method is used to prove each of the following statements? State your answer in such a way that the words "no" and "not" do not appear.

a. There is a real number x such that $x = 2^{-x}$.

b. A implies NOT B.

c. A implies (B implies C).

d. If u is a least upper bound for a set S of real numbers, then $\forall$ real number $\epsilon > 0$, $\exists$ an element $x \in S$ such that $x > u - \epsilon$.

^{B}**10.5**   What statement(s) will you work forward from and what statement(s) will you work backward from if the contrapositive method is used to prove each of the following propositions?

a. (A AND C) implies B.

b. (A OR C) implies B.

c. Suppose that m and n are integers. If n is an even integer and m is an odd integer, then either mn is divisible by 4 or n is not divisible by 4.

10.6   What statement(s) will you work forward from and what statement(s) will you work backward from if the contrapositive method is used to prove each of the following propositions?

a. A implies (C AND D).

b. A implies (C OR D).

c. Suppose that m and n are integers. If either mn is divisible by 4 or n is not divisible by 4, then n is an even integer and m is an odd integer.

^{W}**10.7**   Prove, by the contrapositive method, that if x is a real number that satisfies the property that, for every real number $\epsilon > 0$, $x \geq -\epsilon$, then $x \geq 0$.

10.8   Prove, by contradiction, that if n is a positive integer such that $n^3 - n - 6 = 0$, then for every positive integer m with $m \neq n$, $m^3 - m - 6 \neq 0$.

^{W}**10.9**   Prove, by contradiction, that if x and y are real numbers such that $x \geq 0$, $y \geq 0$, and $x + y = 0$, then $x = 0$ and $y = 0$.

10.10   Prove, by the contrapositive method, that if u is a least upper bound for a set S of real numbers, then $\forall$ real number $\epsilon > 0$, $\exists$ an element $x \in S$ such that $x > u - \epsilon$.

11

Uniqueness Methods and Induction

You now have three major techniques to help you prove that "A implies B": the forward-backward, the contrapositive, and the contradiction methods. In addition, when B has quantifiers, you have the choose and construction methods. Two other special quantifier techniques are developed in this chapter.

11.1 UNIQUENESS METHODS

A **uniqueness method** is used when you want to show not only that there is an object with a certain property such that something happens, but also that the object is unique (that is, there is only one such object). You will know to use the uniqueness method when the statement B contains the key word "unique" (or equivalent words, such as "one and only one," for example) as well as the quantifier "there is." In such a case, your first job is to show that the desired object does exist. This is done either by the construction or the contradiction method. The next step is to show uniqueness in one of two standard ways.

The Direct Uniqueness Method

With the **direct uniqueness method**, you assume that there are two objects having the certain property and for which the something happens. If there really is only one such object then, using the two objects with their certain properties, the something that happens, and the information in A, you must

105

conclude that the two objects are one and the same—that is, that they are equal. The forward-backward method is usually the best way to prove that the two objects are equal. This process is illustrated now.

Proposition 17 *If a, b, c, d, e, and f are real numbers such that $ad - bc \neq 0$, then there are unique real numbers x and y such that $ax + by = e$ and $cx + dy = f$.*

Analysis of Proof.   The existence of the real numbers x and y is proved in Proposition 4 on page 37 via the construction method. Here, the uniqueness is established by the direct uniqueness method. Accordingly, you assume that (x_1, y_1) and (x_2, y_2) are two objects with the certain property and for which the something happens. In this case, that means that

> **A1:** $ax_1 + by_1 = e$ and $cx_1 + dy_1 = f$ and
> **A2:** $ax_2 + by_2 = e$ and $cx_2 + dy_2 = f$.

Using these four equations and the assumption that A is true, the forward-backward method is used to show that the two objects are the same, that is,

> **B1:** $(x_1, y_1) = (x_2, y_2)$.

A key question associated with $B1$ is, "How can I show that two pairs of real numbers are equal?" Using Definition 4 on page 24 for equality of ordered pairs, one answer is to show that

> **B2:** $x_1 = x_2$ and $y_1 = y_2$,

or equivalently, that

> **B3:** $x_1 - x_2 = 0$ and $y_1 - y_2 = 0$.

Both of these statements are obtained from the forward process by applying algebraic manipulations to the four equations in $A1$ and $A2$ and by using the fact that $ad - bc \neq 0$, as indicated in the condensed proof that follows.

Proof of Proposition 17.   The existence of the real numbers x and y is established in Proposition 4 on page 37 via the construction method. Hence only the issue of uniqueness is addressed. To that end, assume that (x_1, y_1) and (x_2, y_2) are real numbers satisfying

1. $ax_1 + by_1 = e$,

2. $cx_1 + dy_1 = f$,

3. $ax_2 + by_2 = e$,

4. $cx_2 + dy_2 = f$.

Subtracting (3) from (1) and (4) from (2) yields

5. $a(x_1 - x_2) + b(y_1 - y_2) = 0$, and

6. $c(x_1 - x_2) + d(y_1 - y_2) = 0$.

On multiplying (5) by d and (6) by b, and then subtracting (6) from (5), it follows that

7. $(ad - bc)(x_1 - x_2) = 0$.

Because, by hypothesis, $ad - bc \neq 0$, one has $x_1 - x_2 = 0$ and hence $x_1 = x_2$. A similar sequence of algebraic manipulations establishes that $y_1 = y_2$ and thus the uniqueness is proved. $\square$

The Indirect Uniqueness Method

With the second method for showing uniqueness, called the **indirect uniqueness method**, you assume that there are two *different* objects having the certain property and for which the something happens. Now supposedly this cannot happen so, by using the certain property, the something that happens, the information in A, and especially the fact that the objects are different, you must reach a contradiction. This process is demonstrated now.

Proposition 18 *If r is a positive real number, then there is a unique real number x such that $x^3 = r$.*

Analysis of Proof. The appearance of the quantifier "there is" in the conclusion suggests using the construction method to produce a real number x such that $x^3 = r$. This part of the proof is omitted and the issue of uniqueness is addressed. To that end, suppose that

   A1: x and y are two different real numbers (that is, $x \neq y$) such that $x^3 = r$ and $y^3 = r$.

Using this information together with the hypothesis that $r > 0$, and especially the fact that $x \neq y$, it is shown that $r = 0$, which contradicts the hypothesis that $r > 0$.

   To show that $r = 0$, work forward from $A1$. Specifically, because $x^3 = r$ and $y^3 = r$, it follows that

   A2: $x^3 = y^3$,   or, $x^3 - y^3 = 0$.

On factoring, one obtains

   A3: $(x - y)(x^2 + xy + y^2) = 0$.

Here is where you can use the fact that $x \neq y$ to divide both sides of $A3$ by $x - y$, obtaining

A4: $x^2 + xy + y^2 = 0$.

Thinking of $A4$ as a quadratic equation of the form $ax^2 + bx + c = 0$, in which $a = 1$, $b = y$, and $c = y^2$, the quadratic formula states that

A5: $x = \frac{-y \pm \sqrt{y^2 - 4y^2}}{2} = \frac{-y \pm \sqrt{-3y^2}}{2}$.

Because x is real and the foregoing formula requires taking the square root of $-3y^2$, it must be that

A6: $y = 0$,

and if $y = 0$, then a contradiction to $r = 0$ is reached from $A1$ because

A7: $r = y^3 = 0$.

Proof of Proposition 18.   Only the issue of uniqueness is addressed. To that end, assume that x and y are two different real numbers for which $x^3 = r$ and $y^3 = r$. Hence it follows that $0 = x^3 - y^3 = (x-y)(x^2 + xy + y^2)$. Because $x \neq y$, it must be that $x^2 + xy + y^2 = 0$. By the quadratic formula,

$$x = \frac{-y \pm \sqrt{y^2 - 4y^2}}{2} = \frac{-y \pm \sqrt{-3y^2}}{2}.$$

Now x is real, so it must be that $y = 0$. But then $r = y^3 = 0$, thus contradicting the hypothesis that $r > 0$, thus completing the proof. □

11.2   INDUCTION

In Chapter 5, you learned to use the choose method when the quantifier "for all" appears in the statement B. There is one special form of B containing the quantifier "for all" for which a separate technique known as **induction** is likely to be more successful.

How to Use Induction

You should consider induction seriously (even before the choose method) when B has the form:

For every integer $n \geq 1$, "something happens,"

where the something that happens is some statement, $P(n)$, that depends on the integer n. The following is an example:

$$\text{For all integers } n \geq 1, \underbrace{\sum_{k=1}^{n} k = \frac{n(n+1)}{2}}_{P(n)}, \quad \text{where } \sum_{k=1}^{n} k = 1 + \cdots + n.$$

When considering induction, the key words to look for are "integer" and "≥ 1."

One way to attempt proving such statements is to make an infinite list of statements, one for each of the integers starting from $n = 1$, and then prove each statement separately. While the first few statements on the list are usually easy to verify, the issue is how to check statement number n and beyond. For the foregoing example, the list is:

$$P(1): \qquad \sum_{k=1}^{1} k = \frac{1(1+1)}{2} \qquad \text{or} \qquad 1 = 1$$

$$P(2): \qquad \sum_{k=1}^{2} k = \frac{2(2+1)}{2} \qquad \text{or} \qquad 1 + 2 = 3$$

$$P(3): \qquad \sum_{k=1}^{3} k = \frac{3(3+1)}{2} \qquad \text{or} \qquad 1 + 2 + 3 = 6$$

$$\vdots$$

$$P(n): \qquad \sum_{k=1}^{n} k = \frac{n(n+1)}{2}$$

$$P(n+1): \qquad \sum_{k=1}^{n+1} k = \frac{(n+1)[(n+1)+1]}{2} \quad = \quad \frac{(n+1)(n+2)}{2}$$

$$\vdots$$

Induction is a clever method for proving that each of these statements in the infinite list is true. Think of induction as a proof machine that starts with $P(1)$ and progresses down the list, proving each statement as it proceeds. Here is how the machine works.

The machine is started by verifying that $P(1)$ is true, which is easy to do for the foregoing example. Then $P(1)$ is fed into the machine. The machine uses the fact that $P(1)$ is true and automatically proves that $P(2)$ is true. You then put $P(2)$ into the machine. The machine uses the fact that $P(2)$ is true to reach the conclusion that $P(3)$ is true, and so on (see Figure 11.1).

Observe that, by the time the machine is going to prove that $P(n+1)$ is true, it will already have shown that $P(n)$ is true (from the previous step). Thus, in designing the machine, you can assume that $P(n)$ is true; your job is to make sure that $P(n+1)$ is also true. In summary, the following steps constitute a proof by induction.

The Steps of Induction

Step 1. Verify that $P(1)$ is true.
Step 2. Use the assumption that $P(n)$ is true to prove that $P(n+1)$ is true.

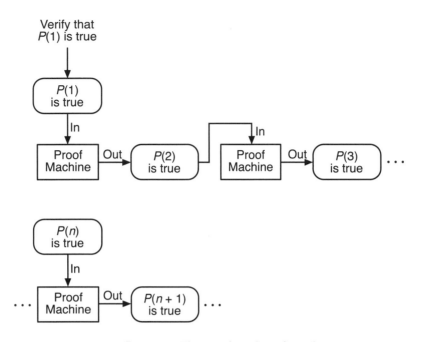

Fig. 11.1   The proof machine for induction.

To perform Step 1, replace n everywhere in $P(n)$ by 1. To verify that the resulting statement is true usually requires only some minor rewriting.

Step 2 is more challenging. You must reach the conclusion that $P(n+1)$ is true by using the assumption that $P(n)$ is true. There is a standard way of doing this. Begin by writing the statement $P(n+1)$, which you want to conclude is true. Because you are assuming that $P(n)$ is true, you should somehow try to rewrite the statement $P(n+1)$ in terms of $P(n)$, for then you can make use of the assumption that $P(n)$ is true. Using the assumption that $P(n)$ is true is referred to as **using the induction hypothesis**. On establishing that $P(n+1)$ is true, the proof is complete. The steps of induction are illustrated with the following proposition.

Proposition 19 *For every integer $n \geq 1$, $\sum_{k=1}^{n} k = \frac{n(n+1)}{2}$.*

Analysis of Proof.   When using the method of induction, it is helpful to write the statement $P(n)$, in this case:

$$P(n): \sum_{k=1}^{n} k = \frac{n(n+1)}{2}.$$

The first step in a proof by induction is to verify $P(1)$. Replacing n everywhere by 1 in $P(n)$, you obtain

$$P(1): \sum_{k=1}^{1} k = \frac{1(1+1)}{2}.$$

With a small amount of rewriting, it is easy to verify this because

$$\sum_{k=1}^{1} k = 1 = \frac{1(1+1)}{2}.$$

This step is often so easy that it is virtually omitted in the condensed proof simply by saying, "The statement is clearly true for $n = 1$."

The second step is more involved. You must use the assumption that $P(n)$ is true to reach the conclusion that $P(n+1)$ is true. The best way to proceed is to write the statement $P(n+1)$ by replacing carefully n everywhere in $P(n)$ with $n + 1$ and rewriting a bit, if necessary. In this case

$$P(n+1): \sum_{k=1}^{n+1} k = \frac{(n+1)[(n+1)+1]}{2} = \frac{(n+1)(n+2)}{2}.$$

To reach the conclusion that $P(n + 1)$ is true, begin with the left side of the equality in $P(n + 1)$ and try to make that side look like the right side. In so doing, you should use the information in $P(n)$ by relating the left side of the equality in $P(n+1)$ to the left side of the equality in $P(n)$. Then you will be able to use the right side of the equality in $P(n)$. In this example,

$$P(n+1): \sum_{k=1}^{n+1} k = \left(\sum_{k=1}^{n} k \right) + (n+1).$$

Now you can use the assumption that $P(n)$ is true by replacing $\sum_{k=1}^{n} k$ with $n(n+1)/2$, obtaining

$$P(n+1): \sum_{k=1}^{n+1} k = \left(\sum_{k=1}^{n} k \right) + (n+1) = \frac{n(n+1)}{2} + (n+1).$$

All that remains is a bit of algebra to rewrite $\frac{n(n+1)}{2} + (n+1)$ as $\frac{(n+1)(n+2)}{2}$, thus obtaining the right side of the equality in $P(n + 1)$. The algebraic steps are:

$$\frac{n(n+1)}{2} + (n+1) = (n+1)\left(\frac{n}{2}+1\right) = \frac{(n+1)(n+2)}{2}.$$

Your ability to relate $P(n+1)$ to $P(n)$ so as to use the induction hypothesis that $P(n)$ is true determines the success of a proof by induction. If you are unable to relate $P(n+1)$ to $P(n)$, then you might wish to consider a different proof technique. On the other hand, if you can relate $P(n + 1)$ to $P(n)$, you will find that induction is easier to use than almost any other technique. To illustrate this fact, you are asked in the exercises to prove Proposition 19 without using induction. Compare your proof with the condensed proof that follows.

Proof of Proposition 19.   The statement is clearly true for $n = 1$. Assume the statement is true for n, that is, that $\sum_{k=1}^{n} k = n(n+1)/2$. Then

$$
\begin{aligned}
\sum_{k=1}^{n+1} k & = \left(\sum_{k=1}^{n} k \right) + (n+1) \\[2mm]
& = \frac{n(n+1)}{2} + (n+1) \\[2mm]
& = (n+1)\left(\frac{n}{2} + 1\right) \\[2mm]
& = \frac{(n+1)(n+2)}{2}.
\end{aligned}
$$

Thus the statement is true for $n+1$ and so the proof is complete. □

Note that induction does not help you to discover the correct form of the statement $P(n)$. Rather, induction only verifies that a given statement $P(n)$ is true for all integers n greater than or equal to some initial one.

Some Variations on Induction

From the foregoing discussion, you know that in the second step in a proof by induction, you use the assumption that $P(n)$ is true to show that $P(n+1)$ is true. From a notational point of view, some authors prefer to use the assumption that $P(n-1)$ is true to show that $P(n)$ is true. These two approaches are identical—either can be used, depending on your notational preference. What is important is that you establish that if a general statement on the infinite list is true, then the next statement is also true.

When using induction, the first value for n need not be 1. For instance, you can use induction to prove that "for all integers $n \geq 5$, $2^n > n^2$." The only modification is that, to start the proof, you must verify $P(n)$ for the first given value of n. In this case, that first value is $n = 5$, so you have to check that $2^5 > 5^2$ (which is true because $2^5 = 32$ while $5^2 = 25$). The second step of the induction proof remains the same—you still have to show that if $P(n)$ is true (that is, $2^n > n^2$), then $P(n+1)$ is also true [that is, $2^{n+1} > (n+1)^2$]. In so doing, you can also use the fact that $n \geq 5$, if necessary.

Another modification to the basic induction method arises when you are having difficulty relating $P(n+1)$ to $P(n)$. Suppose, however, that you can relate $P(n+1)$ to $P(j)$, where $j < n$. In this case, you would like to use the fact that $P(j)$ is true but, can you assume that $P(j)$ is, in fact, true? The answer is yes. To see why, recall the analogy of the proof machine (look again at Figure 11.1) and observe that, by the time the machine has to show that $P(n+1)$ is true, the machine has already proved that all of the statements $P(1), \ldots, P(j), \ldots, P(n)$ are true. Thus, when trying to show that $P(n+1)$ is true, you can assume that $P(n)$ and all preceding statements are true. Such a proof is referred to as **generalized induction** and is illustrated now.

11.3 READING A PROOF

The process of reading and understanding a proof is demonstrated with the following proposition.

Proposition 20 *Any integer $n \geq 2$ can be expressed as a finite product of primes (see Definition 2 on page 24).*

Proof of Proposition 20. (For reference purposes, each sentence of the proof is written on a separate line.)

> **S1:** The statement is clearly true for $n = 2$.
>
> **S2:** Now assume the statement is true for all integers between 2 and n, that is, that any integer j with $2 \leq j \leq n$ can be expressed as a finite product of primes.
>
> **S3:** If $n + 1$ is prime, the statement is true for $n + 1$.
>
> **S4:** Otherwise, $n + 1$ has a prime divisor, that is, there are integers p and q with p prime and $2 \leq q \leq n$ such that $n + 1 = pq$.
>
> **S5:** But by the induction hypothesis, q can be expressed as a finite product of primes and, therefore, so can $n + 1$.

The proof is now complete. □

Analysis of Proof. An interpretation of statements $S1$ through $S5$ follows.

Interpretation of S1: *The statement is clearly true for $n = 2$.*
 The author is performing the first step of induction by mentioning that the statement is true for the first value of n, namely, $n = 2$. The statement is true because 2 is itself prime.

Interpretation of S2: *Now assume the statement is true for all integers between 2 and n, that is, that any integer j with $2 \leq j \leq n$ can be expressed as a finite product of primes.*
 The author is performing the second step of generalized induction by assuming that the statement is true for all integers between 2 and n. It remains to show that the statement is true for $n + 1$.

Interpretation of S3: *If $n + 1$ is prime, the statement is true for $n + 1$.*
 The author notes that the statement is true for $n + 1$ when $n + 1$ is prime, which is clearly correct. Presumably, the author will also show that the statement is true when $n + 1$ is not prime.

Interpretation of S4: *Otherwise, $n + 1$ has a prime divisor, that is, there are integers p and q with p prime and $2 \leq q \leq n$ such that $n + 1 = pq$.*
 The author is showing that the statement is true when $n + 1$ is not prime. Specifically, the author is using the fact that when $n + 1$ is not prime, $n + 1$

has a prime divisor, p. The author then works forward from the fact that p divides $n + 1$ using Definition 1 on 24 to claim that there is an integer q with $2 \leq q \leq n$ such that $n + 1 = pq$.

Interpretation of S5: *But by the induction hypothesis, q can be expressed as a finite product of primes and, therefore, so can $n + 1$.*

The author is applying the induction hypothesis to q, which is valid because q is an integer between 2 and n (see *S4*). Doing so yields that q is a finite product of primes. The author then notes that, as a result, $n + 1 = pq$ is also a finite product of the prime p and the product of primes that constitute q. The generalized induction proof is now complete because the author has correctly established that the statement is true for $n + 1$.

Summary

In this chapter, you have learned two special quantifier techniques: the uniqueness methods and induction.

Uniqueness Methods

Use a uniqueness method when, in the backward process, you come across the need to show that there is a unique object with a certain property such that something happens. With the direct uniqueness method, you

1. Use the construction or contradiction method to establish that there is an object, say, X, with the certain property and for which the something happens.

2. Assume that Y is also an object with the certain property and for which the something happens.

3. Use the properties of X and Y together with the hypothesis A to show that X and Y are the same (that is, that $X = Y$).

With the indirect uniqueness method, you

1. Use the construction or contradiction method to establish that there is an object, say, X, with the certain property and for which the something happens.

2. Assume that Y is a different object from X with the certain property and for which the something happens.

3. Use the properties of X and Y, the fact that they are different, and the hypothesis A to reach a contradiction.

Induction

Use induction when the statement you are trying to prove has the form, "For every integer $n \geq n_0$, $P(n)$," where $P(n)$ is some statement that depends on n. To apply the method of induction,

1. Verify that the statement $P(n)$ is true for n_0. (To do this, replace n everywhere in $P(n)$ by n_0, rewrite the resulting statement, and try to establish that $P(n_0)$ is true.)

2. Assume that $P(n)$ is true.

3. Write the statement $P(n+1)$ by replacing n everywhere in $P(n)$ with $n+1$. (Some rewriting may be necessary to express $P(n+1)$ in a clean form.)

4. Reach the conclusion that $P(n+1)$ is true. To do so, relate $P(n+1)$ to $P(n)$ and then use the fact that $P(n)$ is true. The key to using induction rests in your ability to relate $P(n+1)$ to $P(n)$.

Exercises

Note: Solutions to exercises marked with a B are in the back of this book. Solutions to exercises marked with a W are located on the World Wide Web at http://www.wiley.com/college/solow/.

Note: All proofs, *except those by induction*, should contain an analysis of proof and a condensed version. Definitions for all mathematical terms are provided in the glossary at the end of the book.

^{B}**11.1**   Prove that if x is a real number > 2, then there is a unique real number $y < 0$ such that $x = 2y/(1+y)$.

11.2   Prove that if a and b are integers with $a \neq 0$ such that $a|b$, then there is a unique integer k such that $b = ka$. (See Definition 1 on page 24.)

^{W}**11.3**   Prove, by the indirect uniqueness method, that if m and b are real numbers with $m \neq 0$, then there is a unique number x such that $mx + b = 0$.

11.4   Prove, by the indirect uniqueness method, that there is a unique integer n for which $2n^2 - 3n - 2 = 0$.

^{B}**11.5**   Prove that if a and b are real numbers, at least one of which is not 0, and $i = \sqrt{-1}$, then there is a unique complex number, say, $c + di$, such that $(a + bi)(c + di) = 1$.

^{B}**11.6**   For which of the following statements is induction applicable? When induction is not applicable, explain why not.

    a. For every positive integer n, 8 divides $5^n + 2(3^{n-1}) + 1$.

b. There is an integer $n \geq 0$ such that $2n > n^2$.

c. For every integer $n \geq 1$, $1(1!) + \cdots + n(n!) = (n+1)! - 1$. (Recall that $n! = n(n-1) \cdots 1$.)

d. For every integer $n \geq 4$, $n! > n^2$.

e. For every real number $n \geq 1$, $n^2 \geq n$.

11.7   Upon learning about the method of induction, a student said, "I do not understand something. After showing that the statement is true for $n = 1$, you want me to assume that $P(n)$ is true and to show that $P(n+1)$ is true. How can I assume that $P(n)$ is true—after all, aren't we trying to *show* that $P(n)$ is true?" Answer this question.

^{W}**11.8**   With regard to induction,

a. Why and when would you want to use induction instead of the choose method?

b. Why is it not possible to use induction on statements of the form, "For every real number with a certain property, something happens"?

11.9   Suppose that $A(n)$ and $B(n)$ are statements that depend on a positive integer n. Explain how you would use induction to prove that for every integer $n \geq 1$, if $A(n)$ is true, then $B(n)$ is true. For Step 2 of induction, indicate what statement(s) you would assume are true and what statement(s) you would have to show are true.

^{B}**11.10**   Prove, by induction, that for every integer $n \geq 1$, $1(1!) + \cdots + n(n!) = (n+1)! - 1$.

11.11   Prove, by induction, that for every integer $k \geq 1$, $1 + 2 + 2^2 + \cdots + 2^{k-1} = 2^k - 1$.

^{W}**11.12**   Prove, by induction, that for every integer $n \geq 5$, $2^n > n^2$.

11.13   Prove, by induction, that for every integer $n \geq 1$,

$$\frac{1}{n!} \leq \frac{1}{2^{n-1}}.$$

^{B}**11.14**   Prove, by induction, that a set of $n \geq 1$ elements has 2^n subsets (including the empty set).

11.15   Prove, by induction, that if $x > 1$ is a given real number, then for every integer $n \geq 2$, $(1+x)^n > 1 + nx$.

^{B}**11.16**   Prove, *without using induction*, that for any integer $n \geq 1$, $\sum_{k=1}^{n} k = n(n+1)/2$.

11.17   A machine is filled with an odd number of chocolate candies and an odd number of caramel candies. For 25 cents, the machine dispenses two candies. Prove that, before being empty, the machine will dispense at least one pair that consists of one chocolate candy and one caramel candy.

^{W}**11.18**   Prove, by induction, that, if $i = \sqrt{-1}$, then for every integer $n \geq 1$, $[\cos(x) + i\sin(x)]^n = \cos(nx) + i\sin(nx)$. Do this by showing that (1) the statement is true for $n = 1$ and (2) if the statement is true for $n - 1$, then the statement is true for n.

11.19   Prove that in a line of at least two people, if the first person is a woman and the last person is a man, then somewhere in the line there is a man standing immediately behind a woman.

^{B}**11.20**   Describe a modified induction procedure for proving statements of the form:

    a. For every integer $n \leq n_0$, $P(n)$ is true.

    b. For every integer n, $P(n)$ is true.

11.21   Describe a modified induction procedure for proving that for every positive odd integer n, $P(n)$ is true.

^{B}**11.22**   In the following condensed proof, explain how the author relates $P(n + 1)$ to $P(n)$ and where the induction hypothesis is used.

> **Proposition.** $\forall$ integer $n \geq 2$, $\displaystyle\prod_{k=2}^{n} (1 - \frac{1}{k^2}) = \frac{n+1}{2n}$.

> **Proof.** The statement is true for $n = 2$ because $1 - \frac{1}{4} = \frac{3}{4} = \frac{2+1}{2(2)}$. Assume the statement is true for n. Then

$$\prod_{k=2}^{n+1} (1 - \tfrac{1}{k^2}) = \left(\prod_{k=2}^{n} (1 - \tfrac{1}{k^2})\right)\left(1 - \tfrac{1}{(n+1)^2}\right)$$

$$= \tfrac{n+1}{2n} - \tfrac{1}{2n(n+1)}$$

$$= \tfrac{n+2}{2(n+1)}.$$

> The proof is now complete. $\square$

^{W}**11.23**   In the following condensed proof, explain how the author relates $P(n + 1)$ to $P(n)$ and where the induction hypothesis is used.

> **Proposition.** For every integer $n \geq 1$, the derivative of x^n is nx^{n-1}.

Proof. The statement is true for $n = 1$ because, $(x)' = 1 = 1x^0$. Assume now that $(x^n)' = nx^{n-1}$. Then, for x^{n+1}, you have that

$$
\begin{aligned}
(x^{n+1})' &= [(x)(x^n)]' \\
&= (x)'x^n + x(x^n)' \\
&= x^n + x(nx^{n-1}) \\
&= (n+1)x^n.
\end{aligned}
$$

The proof is now complete. $\square$

11.24 In the following condensed proof, explain how the author relates $P(n+1)$ to $P(n)$ and where the induction hypothesis is used.

Proposition. For every integer $n \geq 2$, if $x_1, x_2, \ldots, x_n$ are real numbers > 0 and < 1, then $(1-x_1)(1-x_2)\cdots(1-x_n) > 1 - x_1 - x_2 - \cdots - x_n$.

Proof. When $n = 2$, the statement is true because you have that $(1-x_1)(1-x_2) = 1-x_1-x_2+x_1x_2 > 1-x_1-x_2$. Assume the statement is true for n. Then,

$$
\begin{aligned}
(1 - x_1)(1 - x_2)&\cdots(1 - x_{n+1}) \\
&= [(1 - x_1)(1 - x_2)\cdots(1 - x_n)](1 - x_{n+1}) \\
&> (1 - x_1 - \cdots - x_n)(1 - x_{n+1}) \\
&= 1 - x_1 - \cdots - x_n - (1 - x_1 - x_2 - \cdots - x_n)(x_{n+1}) \\
&= 1 - x_1 - \cdots - x_n - x_{n+1} + (x_1 + x_2 + \cdots + x_n)(x_{n+1}) \\
&> 1 - x_1 - \cdots - x_n - x_{n+1}.
\end{aligned}
$$

The proof is now complete. $\square$

[B]**11.25** What, if anything, is wrong with the following proof that all horses have the same color?

Proof. Let n be the number of horses. When $n = 1$, the statement is clearly true, that is, one horse has the same color, whatever color it is. Assume that any group of n horses has the same color. Now consider a group of $n + 1$ horses. Taking any n of them, the induction hypothesis states that they all have the same color, say, brown. The only issue is the color of the remaining "uncolored" horse. Consider, therefore, any other group of n of the $n+1$ horses that contains the uncolored horse. Again, by the induction hypothesis, all of the horses in the new group must have the same color. Then, because all of the colored horses in this group are brown, the uncolored horse must also be brown. $\square$

^{W} **11.26**   What, if anything, is wrong with the following condensed proof?

Proposition. If r is a real number with $|r| \leq 1$, then for all integers $n \geq 1$, $1 + r + r^2 + \cdots + r^{n-1} = (1 - r^n)/(1 - r)$.

Proof. The statement is clearly true for $n = 1$. Assume it is true for n. Then, for $n + 1$, one has,

$$
\begin{aligned}
1 + \cdots + r^n &= (1 - r^n)/(1 - r) + r^n \\
&= (1 - r^n + r^n - r^{n+1})/(1 - r) \\
&= (1 - r^{n+1})/(1 - r).
\end{aligned}
$$

The proof is now complete.   □

11.27   What, if anything, is wrong with the following condensed proof?

Proposition. For every integer $n \geq 2$, if $S_1, S_2, \ldots, S_n$ are convex sets of real numbers, then $S_1 \cup S_2 \cup \cdots \cup S_n$ is a convex set.

Proof. To see that the statement is true for $n = 2$, let $x, y \in S_1 \cup S_2$ and let t be a real number with $0 \leq t \leq 1$. Because $x, y \in S_1 \cup S_2$ and because S_1 is convex, $tx + (1 - t)y \in S_1$. Likewise, because S_2 is convex, $tx + (1 - t)y \in S_2$. Thus, $tx + (1 - t)y \in S_1 \cup S_2$ and so the union of two convex sets is a convex set. Assume now that the statement is true for n and let $S_1, S_2, \ldots, S_{n+1}$ be convex sets of real numbers. You then have that

$$S_1 \cup \cdots \cup S_{n+1} = (S_1 \cup \cdots \cup S_n) \cup S_{n+1}.$$

By the induction hypothesis, $S = S_1 \cup \cdots \cup S_n$ is convex. Finally, it has already been shown that the union of two convex sets is convex and therefore $S \cup S_{n+1}$ is convex, completing the proof.   □

12

Either/Or and Max/Min Methods

Two final techniques are presented in this chapter. The first is associated with the key words "either/or" and the second with the key words "max/min."

12.1 EITHER/OR METHODS

The **either/or methods** arise when you come across the key words "either/or" in the form "either C or D is true," (where C and D are statements). Two different techniques are available depending, respectively, on whether these key words arise in the forward or backward process.

Proof by Elimination

To illustrate one of these methods, called a **proof by elimination**, suppose that the key words "either/or" arise in the conclusion of a proposition whose form is "A implies C OR D." With the forward-backward method, you assume A is true and need to conclude that either C is true or else D is true. The only question is whether you should show that C is true or whether you should show that D is true. To eliminate the uncertainty, suppose you were to make the additional assumption that C is not true. Clearly it had better turn out that, in this case, D is true. Thus, with a proof by elimination, you assume that A is true and C is false; you must then conclude that D is true, as is illustrated now.

Proposition 21 *If $x^2 - 5x + 6 \geq 0$, then $x \leq 2$ or $x \geq 3$.*

Analysis of Proof.   Recognizing the key words "either/or" in the conclusion, you should proceed with a proof by elimination. Accordingly, assume that

A: $x^2 - 5x + 6 \leq 0,$ and
A1 (*NOT C*): $x > 2.$

It is your job to conclude that

B1 (*D*): $x \geq 3.$

Working forward from A by factoring, it follows that

A2: $(x - 2)(x - 3) \geq 0.$

Because $x > 2$ from $A1$, you can divide both sides of $A2$ by the positive number $x - 2$ to obtain

A3: $x - 3 \geq 0.$

Adding 3 to both sides of $A3$ yields $B1$, thus completing the proof.

Proof of Proposition 21.   Assume that $x^2 - 5x + 6 \geq 0$ and $x > 2$. It follows that $(x - 2)(x - 3) \geq 0$. Because $x > 2$, $x - 2 > 0$, and so it must be that $x \geq 3$, as desired. □

Note that you can do a proof by elimination of "A implies C OR D" equally well by assuming that A is true and D is false and then concluding that C is true. Try this approach on Proposition 21.

Proof by Cases

The second either/or method, called a **proof by cases**, is used when the hypothesis contains the key words either/or in the form, "C OR D implies B." According to the forward-backward method, you can assume that C OR D is true; you must conclude that B is true. The only question is, "Is it C that is true or is it D that is true?" Because you do not know which of these is true, you should proceed by cases, that is, you should do two proofs. In the first one you assume that C is true and prove that B is true; in the second one you assume that D is true and prove that B is true. This technique is illustrated now. Observe how the words either/or are introduced intentionally in the middle of the proof so as to use a proof by cases.

Proposition 22 *If a is a negative real number, then $y = -b/(2a)$ is a maximum of the function $ax^2 + bx + c$.*

Analysis of Proof.   Because no key words appear in the hypothesis or conclusion, the forward-backward method is used to begin the proof. Working backward, the key question is, "How can I show that a point [namely, $y =$

$-b/(2a)]$ is a maximum of a function?" Applying the definition given in Exercise 5.1(a) on page 54, it is necessary to show that

B1: For every real number x, $ay^2 + by + c \geq ax^2 + bx + c$.

Recognizing the key words "for all" in the backward process, the choose method is used to choose

A1: A real number x,

for which it must be shown that

B2: $ay^2 + by + c \geq ax^2 + bx + c$.

Subtracting $ax^2 + bx + c$ from both sides of B2 and factoring out $y - x$, it must be shown that

B3: $(y - x)[a(y + x) + b] \geq 0$.

If $y - x = 0$, then B3 is true. Thus you can assume that

A2: $y - x \neq 0$.

It is important to note here that you can rewrite A2 so as to contain the key words "either/or" explicitly:

A3: Either $y - x > 0$ or $y - x < 0$.

At this point, recognizing the key words "either/or" in the forward process, it is time to use a proof by cases. Accordingly, you should do two proofs—first assume that $y - x > 0$ and prove that B3 is true; then assume that $y - x < 0$ and again prove that B3 is true. These two proofs are done now.

Case 1: Assume that

A4: $y - x > 0$.

In this case you can divide both sides of B3 by the positive number $y - x$; thus it must be shown that

B4: $a(y + x) + b \geq 0$.

Working forward from the fact that $y = -b/(2a)$ and $a < 0$ (see the hypothesis), it follows from A4 that

A5: $2ax + b > 0$

and so

A6: $a(y + x) + b = ax + b/2 = (2ax + b)/2 > 0$.

Thus B3 is true and this completes the first case.

Case 2: Now assume that

A4: $y - x < 0$.

In this case you can divide both sides of $B3$ by the negative number $y - x$; thus it is necessary to show that

B4: $a(y + x) + b \leq 0$.

Working forward from the fact that $y = -b/(2a)$ and $a < 0$ (see the hypothesis), it follows from $A4$ that

A5: $2ax + b < 0$

and so

A6: $a(y + x) + b = ax + b/2 = (2ax + b)/2 < 0$.

Thus $B4$ is true and this completes the second case and, in fact, the entire proof.

Observe that the proof of the foregoing second case is almost identical to that of the first case except for a reversal of sign in several places. Most mathematicians would not, in general, write the details for both cases. Rather, when they recognize the similarity of the two cases, they would say, "Assume, without loss of generality, that Case 1 occurs ...." They would then proceed to present the details for that case, omitting the second case altogether. In other words, the words "assume, without loss of generality, ..." mean that the author will present only one of the cases in detail; you will have to provide the details of the other case for yourself, as is illustrated in the following condensed proof of Proposition 22.

Proof of Proposition 22.   Let x be a real number. (The word "let" indicates that the choose method is used.) It is shown that $ay^2 + by + c \geq ax^2 + bx + c$ or, equivalently, that $(y - x)[a(y + x) + b] \geq 0$. This is clearly true if $y - x = 0$, so assume that $y - x \neq 0$. Then either $y - x > 0$ or $y - x < 0$. Assume, without loss of generality, that $y - x > 0$. Because $a < 0$ and $y = -b/(2a)$, it follows that $[a(y + x) + b] > 0$, and so the proof is complete. $\square$

Observe that with a proof by cases, two separate proofs are needed to show that "C OR D implies B"—you must prove both that "C implies B" and then that "D implies B." In contrast, with a proof by elimination, only one proof is needed to show that "A implies C OR D"—you can prove either "A AND (NOT C) implies D," or "A AND (NOT D) implies C"; either one of these two proofs by itself suffices.

12.2   READING A PROOF

The process of reading and understanding a proof is demonstrated with the following proposition.

Proposition 23 *If p and b are positive integers for which p is prime and p does not divide b, then the only positive integer that divides both p and b is 1.*

Proof of Proposition 23.   (For reference purposes, each sentence of the proof is written on a separate line.)

S1:   Clearly 1 divides both p and b.
S2:   To see that 1 is the only such integer, let $d > 0$ be an integer that divides both b and p.
S3:   Thus, there is an integer k such that $b = kd$.
S4:   Because p is prime and d divides p, it must be that $d = 1$ or $d = p$.
S5:   Now $d \neq p$, for otherwise $b = kp$ and p would divide b.
S6:   It therefore follows that $d = 1$.

The proof is now complete.   □

Analysis of Proof.   An interpretation of statements $S1$ through $S6$ follows.

Interpretation of S1: *Clearly 1 divides both p and b.*
    The author recognizes the key words "one and only one" in the conclusion and therefore uses a uniqueness method. Accordingly, the author first states that 1 divides both p and b (which is true because 1 divides every integer).

Interpretation of S2: *To see that 1 is the only such integer, let $d > 0$ be an integer that divides both b and p.*
    This statement indicates that the author is using the direct uniqueness method and therefore assumes that

A1: $d > 0$ divides both b and p.

To complete the direct uniqueness method, the author must show that the two objects, d and 1, are the same, that is, that

B1: $d = 1$.

Indeed the author reaches this conclusion in $S6$.

Interpretation of S3: *Thus, there is an integer k such that b = kd.*
    The author is working forward by definition from the fact that d divides b (see $A1$) to claim that

A2: There is an integer k such that $b = kd$.

Interpretation of S4: *Because p is prime and d divides p, it must be that* $d = 1$ *or* $d = p$.

The author is working forward by definition from the hypothesis that p is prime, so,

  A3: The only positive integers that divide p are 1 and p.

The author rewords $A3$ to read

  A4: For every positive integer a that divides p, $a = 1$ or $a = p$.

In this form, the author recognizes the quantifier "for all" in the forward process and specializes $A4$ to $a = d > 0$, which does divide p (see $A1$). The result of this specialization, as the author claims in $S4$, is

  A5: $d = 1$ or $d = p$.

Interpretation of S5: *Now* $d \neq p$, *for otherwise* $b = kp$ *and* p *would divide* b.

Recognizing the key words "either/or" in $A5$, the author proceeds with a proof by cases and assumes first that $d = p$. However, the author shows by contradiction that this is impossible. Specifically, the author works forward by substituting $d = p$ into $A2$, which results in $b = kp$. By definition, this means that p divides b, which contradicts the hypothesis that p does not divide d.

Interpretation of S6: *It therefore follows that* $d = 1$.

Having ruled out the first case of $d = p$ in $A5$, the author now proceeds to the second case, which is that $d = 1$, as stated in $S6$. However, the fact that $d = 1$ completes the direct uniqueness method (see $B1$) and hence the proof.

12.3 MAX/MIN METHODS

The final techniques presented in this chapter are the **max/min methods** that arise in problems dealing with maxima and minima. Suppose that S is a nonempty set of real numbers having both a largest and a smallest member. For a given real number x, you might be interested in the position of the set S relative to the number x. For instance, you might want to prove one of the following statements:

1. All of S is to the right of x [see Figure 12.1(a)].

2. Some of S is to the left of x [see Figure 12.1(b)].

3. All of S is to the left of x [see Figure 12.1(c)].

4. Some of S is to the right of x [see Figure 12.1(d)].

In mathematical problems these four statements are likely to appear, respectively, as:

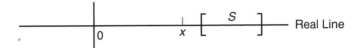

Figure 12.1 (a) – All of S to the right of x.

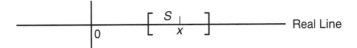

Figure 12.1 (b) – Some of S to the left of x.

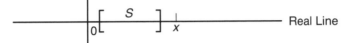

Figure 12.1 (c) – All of S to the left of x.

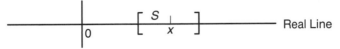

Figure 12.1 (d) – Some of S to the right of x.

Fig. 12.1   Possible positions of a set S relative to a real number x.

(a) $\min\{s : s \in S\} \geq x$.

(b) $\min\{s : s \in S\} \leq x$.

(c) $\max\{s : s \in S\} \leq x$.

(d) $\max\{s : s \in S\} \geq x$.

The technique associated with the first two are discussed here, and the remaining two are left as exercises. The idea behind the max/min techniques is to convert the given statement into an equivalent one containing a quantifier; then you can use the appropriate choose, construction, or specialization method.

Consider, therefore, the problem of trying to show that the smallest member of S is $\geq x$. An equivalent problem containing a quantifier is obtained by considering the foregoing statement 1. Because *all* of S should be to the right of x, you need to show that, for all elements $s \in S$, $s \geq x$, as is illustrated now.

Proposition 24 *If R is the set of all real numbers, then* $\min\{x(x - 2) : x \in R\} \geq -1$.

Analysis of Proof. From the form of B, the max/min method is used. As per the foregoing discussion, you can convert the statement B to

B1: For all real numbers x, $x(x - 2) \geq -1$.

Once in this form, it is clear that you should use the choose method to choose

A1: A real number x,

for which it must be shown that

B2: $x(x - 2) \geq -1$.

Rewriting $B2$ means you must show that

B3: $x^2 - 2x + 1 \geq 0$,

or equivalently, that

B4: $(x - 1)^2 \geq 0$.

This, however, is always true, so the proof is complete.

Proof of Proposition 24. To conclude that $\min\{x(x - 2) : x \in R\} \geq -1$, let x be any real number. Then it follows that $x(x - 2) \geq -1$ because $x^2 - 2x + 1 = (x - 1)^2 \geq 0$. □

Turning now to the problem of showing that the smallest member of S is $\leq x$, the approach is slightly different. To proceed, consider the foregoing statement 2. Because *some* of S should be to the left of x, an equivalent problem is to show that there is an element $s \in S$ such that $s \leq x$. The construction or contradiction method is the used to do so.

Summary

You have now learned two techniques that are appropriate when statements have the key words "either/or" and "max/min."

Either/Or Methods

Use an either/or method when you encounter these key words in the forward or backward process. A proof by elimination is used to show that "A implies C OR D." To do so, follow these steps.

1. Assume that A and $NOT\ C$ are true.

2. Work forward from A and $NOT\ C$ to establish that D is true.

3. Work backward from D.

(You could equally well assume that A and $NOT\ D$ are true, and work forward to prove that C is true. You can also work backward from C in this case.)

Use a proof by cases to show that "$C\ OR\ D$ implies B." To do so you must do two proofs, that is,

Case 1: Prove that C implies B.

Case 2: Prove that D implies B.

Max/Min Methods

Use a max/min method when you need to show that the largest or smallest element of a set is $\leq$ or $\geq$ some fixed number. To do so, convert the statement into an equivalent statement containing a quantifier and then apply the choose, construction, or specialization method, whichever is appropriate.

Exercises

Note: Solutions to exercises marked with a B are in the back of this book. Solutions to exercises marked with a W are located on the World Wide Web at http://www.wiley.com/college/solow/.

Note: All proofs should contain an analysis of proof and a condensed version. Definitions for all mathematical terms are provided in the glossary at the end of the book.

^{B}**12.1**   Answer each of the following questions pertaining to the either/or methods.

   a. Describe how a proof by elimination is applied to prove a statement of the form, "If A, then $C\ OR\ D\ OR\ E$."

   b. Describe how a proof by cases is applied to prove a statement of the form, "If $C\ OR\ D\ OR\ E$, then B."

12.2   Assuming you use the contradiction method on each of the following problems, what technique will you use to work forward from $NOT\ B$? Explain.

   a. A implies $(C\ AND\ D)$.

   b. A implies $[(NOT\ C)\ AND\ (NOT\ D)]$.

^{B}**12.3**   Explain where, why, and how a proof by cases is used in the proof of Proposition 20 on page 113.

12.4   Explain where, why, and how a proof by cases is used in the condensed proof for Proposition 6 on 51.

^{W}**12.5**   Explain where, why, and how a proof by cases is used in the condensed proof presented in Exercise 8.19 on page 89.

12.6   Explain where, why, and how a proof by elimination is used in the following proof.

> **Proposition.** If $i = \sqrt{-1}$ and $a + bi$ and $c + di$ are complex numbers for which $(a + bi)(c + di) = 1$, then $a \neq 0$ or $b \neq 0$.
>
> **Proof.** Because $(a + bi)(c + di) = 1$, it follows that $ac + adi + bci - bd = 1$. That is, $ac - bd = 1$ and $ad + bc = 0$. If $a = 0$, then $-bd = 1$. It then follows that $b \neq 0$, thus completing the proof. $\square$

^{B}**12.7**   Consider the proposition, "If x is a real number that satisfies $x^3 + 3x^2 - 9x - 27 \geq 0$, then $|x| \geq 3$."

a. Reword the proposition so that it is of the form "A implies C OR D."

b. Prove the proposition by assuming that A and $NOT\ C$ are true.

^{W}**12.8**   Prove the proposition in the previous exercise by assuming that A and $NOT\ D$ are true.

^{W}**12.9**   Prove that if a and b are integers for which $a|b$ and $b|a$, then $a = \pm b$. (See Definition 1 on page 24.)

12.10   Provide an analysis of proof for the following condensed proof.

> **Proposition.** If n is a positive integer, then either n is prime, or n is a square, or n divides $(n - 1)!$.
>
> **Proof.** If $n = 1$, then $n = 1^2$ is a square and the proposition is true. Similarly, if $n = 2$, then n is prime and again the proposition is true. So suppose that $n > 2$ is neither prime nor a square. Because $n > 2$ is not prime, there are integers a and b with $1 < a < n$ and $1 < b < n$ such that $n = ab$. Also, because n is not square, $a \neq b$. This means that a and b are integers with $2 \leq a \neq b \leq n - 1$. That is, a and b are two different terms of $(n - 1)(n - 2) \cdots 1$. Thus, $ab = n$ divides $(n - 1)!$. $\square$

^{B}**12.11**   Prove that if a, b, and c are integers for which either $a|b$ or $a|c$, then $a|(bc)$.

12.12   Prove that if S and T are subsets of a universal set U, then $(S \cap T)^c = S^c \cup T^c$, where $X^c = \{x \in U : x \notin X\}$. Indicate clearly where the either/or methods arise.

^{B}**12.13**   Convert the following max/min problems to an equivalent statement containing a quantifier. (Note: S is a set of real numbers and x is a given real number.)

    a. $\max\{s : s \in S\} \leq x$.

    b. $\max\{s : s \in S\} \geq x$.

In the remainder of this problem, a, b, c, and u are given real numbers, and x is a variable.

    c. $\min\{cx : ax \leq b$ and $x \geq 0\} \leq u$.

    d. $\max\{cx : ax \leq b$ and $x \geq 0\} \geq u$.

    e. $\min\{ax : b \leq x \leq c\} \geq u$.

    f. $\max\{ax : b \leq x \leq c\} \leq u$.

12.14   Convert each of the following statements into an equivalent statement having a quantifier.

    a. The maximum of a function $f(x) \leq y$, where x is a real number with $0 \leq x \leq 1$ and y is a given real number.

    b. The minimum of a function $f(x) \leq y$, where x is a real number with $0 \leq x \leq 1$ and y is a given real number.

^{B}**12.15**   Suppose that a, b, and c are given real numbers and that x and u are variables. Prove that $\min\{cx : ax \geq b, x \geq 0\}$ is at least as large as $\max\{ub : ua \leq c, u \geq 0\}$.

^{W}**12.16**   Prove that if S is a nonempty subset of a set T of real numbers and t^* is a real number such that for each element $t \in T$, $t \geq t^*$, then $\min\{s : s \in S\} \geq t^*$.

13

Summary

The list of proof techniques is now complete. The techniques presented here are not the only ones, but they do constitute the basic set. You will come across others as you are exposed to more mathematics—perhaps you will develop some of your own. In any event, there are many fine points and tricks that you will pick up with experience. A final summary of how and when to use each of the various techniques for proving the proposition "A implies B" is in order.

13.1 THE FORWARD-BACKWARD METHOD

With the forward-backward method, you assume that A is true and your job is to prove that B is true. Through the forward process, you derive from A a sequence of statements, $A1$, $A2$, ..., that are necessarily true as a result of assuming that A is true. This sequence is guided by the backward process whereby, through asking and answering the key question, you derive from B a new statement, $B1$, with the property that if $B1$ is true, then so is B. This backward process is then applied to $B1$, resulting in a new statement, $B2$, and so on. The objective is to link the forward sequence to the backward sequence by generating a statement in the forward sequence that is the same as the last statement obtained in the backward sequence. Then, like a column of dominoes, you can do the proof by going forward along the sequence from A all the way to B.

13.2   THE CONSTRUCTION METHOD

When obtaining the sequence of statements, watch for quantifiers to appear, for then the construction, choose, and/or specialization methods are likely to be useful in doing the proof. For instance, when the quantifier "there is" arises in the backward process in the standard form:

> There is an "object" with a "certain property" such that "something happens,"

consider using the construction method to produce the desired object. With the construction method, you work forward from the assumption that A is true to construct (produce, or devise an algorithm to produce, and so on) the object. However, the actual proof consists of showing that the object you construct satisfies the certain property and also that the something happens.

13.3   THE CHOOSE METHOD

On the other hand, when the quantifier "for all" arises in the backward process in the standard form:

> For all "objects" with a "certain property," "something happens,"

consider using the choose method. Here, your objective is to design a model proof for establishing that the something happens for a general object with the certain property. If successful, then you could, in theory, repeat this proof for each specific object with the certain property. You therefore select (or choose) an object that has the certain property. You must conclude that, for the chosen object, the something happens. Once you choose the object, work forward from the fact that the chosen object does have the certain property (together with the information in A) and backward from the something that happens.

13.4   THE SPECIALIZATION METHOD

When the quantifier "for all" arises in the forward process in the standard form:
> For all "objects" with a "certain property," "something happens,"

you will probably want to use the specialization method. To do so, watch for one of these objects to arise, often in the backward process. By using specialization, you can then conclude, as a new statement in the forward process, that the something does happen for that particular object. That fact should then be helpful in reaching the conclusion that B is true. When using specialization, be sure to verify that the particular object does satisfy the certain property, for only then does the something happen.

If statements contain more than one quantifier, that is, they are nested, process them in the order in which they appear from left to right. As you read the first quantifier in the statement, identify its objects, certain property, and something that happens. Then apply an appropriate technique based on whether the statement is in the forward or backward process, and whether the quantifier is "for all" or "there is." This process is repeated until all quantifiers are dealt with.

13.5   THE CONTRADICTION METHOD

When the statement B contains the key word "no" or "not," or when the forward-backward method fails, you should consider the contradiction method. With this approach, you assume not only that A is true but also that B is false. This gives you two facts from which you must derive a contradiction to something that you know to be true. Where the contradiction arises is not always obvious but the contradiction is obtained by working forward from the statements A and $NOT\ B$.

13.6   THE CONTRAPOSITIVE METHOD

In the event that the contradiction method fails, there is still hope with the contrapositive method. To use the contrapositive approach, write the statements $NOT\ B$ and $NOT\ A$ using the techniques of Chapter 10. Then, by beginning with the assumption that $NOT\ B$ is true, your job is to conclude that $NOT\ A$ is true. This is best accomplished by applying the forward-backward method, working forward from $NOT\ B$ and backward from $NOT\ A$. Remember to watch for quantifiers to appear in the forward and backward processes for, if they do, then the corresponding construction, choose, and/or specialization methods may be useful.

13.7   THE UNIQUENESS METHODS

When the conclusion of a proposition requires you to show that

> There is a unique "object" with a "certain property" such
> that "something happens,"

you can use the direct uniqueness method. With this method, you first establish the existence of the desired object, say, X, using the construction (or contradiction) method. Then you assume that Y is a second object satisfying the certain property and for which the something happens. Working forward from all this information (together with the hypothesis), you must conclude that the two objects, X and Y, are the same, that is, that they are equal. You can also work backward from the fact that X is the same as Y.

With the indirect uniqueness method, you also begin by establishing the existence of the desired object X. To show uniqueness, however, you then assume that Y is a different object that also satisfies the certain property and the something that happens. You must use all this information—especially the fact that X and Y are different—to reach a contradiction.

13.8   THE INDUCTION METHOD

Consider the induction method (even before the choose method) when the statement B has the form:

> For every integer n greater than or equal to some initial integer, a statement $P(n)$ is true.

The first step of the induction method is to verify that $P(n)$ is true for the first possible value of n. The second step requires you to show that if $P(n)$ is true, then $P(n+1)$ is true. The success of a proof by induction rests on your ability to relate $P(n+1)$ to $P(n)$ so that you can use the assumption that $P(n)$ is true. In other words, to perform the second step of the induction proof, write the statement $P(n)$, replace n everywhere by $n+1$ to obtain $P(n+1)$, and then see if you can express $P(n+1)$ in terms of $P(n)$. Only then will you be able to use the assumption that $P(n)$ is true to reach the conclusion that $P(n+1)$ is also true.

13.9   EITHER/OR METHODS

When the key words "either/or" arise, there are two proof techniques available depending, respectively, on whether those key words appear in the forward or backward processes. For example, you should use a proof by elimination when trying to prove a proposition of the form, "If A, then C OR D." To do so, assume that A is true and C is not true (that is, A and NOT C); you should then show that D is true. This is best accomplished with the forward-backward method. Alternatively, you can assume that A and NOT D are true; in this case you would have to show that C is true.

A proof by cases is used when trying to prove a proposition of the form, "If C OR D, then B." Two proofs are required. In the first case, you assume that C is true and then prove that B is true; in the second case, you assume that D is true and then prove that B is true.

13.10   THE MAX/MIN METHODS

When the conclusion of a proposition requires you to show that the smallest (largest) element of a set of real numbers is less (greater) than or equal to a particular real number, say, x, then you should use a max/min method. Doing so involves rewriting the statement in an equivalent form using the quantifier "for all" or "there is," whichever is appropriate. Once in this form, you can then apply the choose, construction, or specialization method.

Conclusion

In trying to prove that "A implies B," let the form of A and B guide you as much as possible. For example, you should scan the statement B for certain key words, which often indicate how to proceed. If you come across the quantifier "there is," then consider the construction method, whereas the quantifier "for all" suggests using the choose or induction method. When the statement B contains the word "no" or "not," you will probably want to use the contrapositive or contradiction method. Other key words to look for are "uniqueness," "either/or," and "maximum" and "minimum," for then the corresponding uniqueness, either/or, and max/min methods are appropriate. If you are unable to choose an approach based on the form of B, then you should proceed with the forward-backward method. Tables 13.1 and 13.2 provide a complete summary.

You are now ready to "speak" mathematics. Your new "vocabulary" and "grammar" are complete. You have learned the three major proof techniques for proving propositions, theorems, lemmas, and corollaries: the forward-backward, contrapositive, and contradiction methods. You have come to know the quantifiers and the corresponding construction, choose, induction, and specialization methods. For special situations, your bag of proof techniques includes the uniqueness, either/or, and max/min methods. If all of these proof techniques fail, you may wish to stick to Greek—after all, it's all Greek to me.

Table 13.1   Summary of Main Proof Techniques

Proof Technique	When to Use It	What to Assume
Forward-Backward (page 9)	As a first attempt, or when B does not have a recognizable form.	A
Contrapositive (page 91)	When B has the word "no" or "not" in it.	$NOT\ B$
Contradiction (page 80)	When B has the word "no" or "not" in it, or when the first two methods fail.	A and $NOT\ B$
Construction (page 37)	When B has the words "there is," "there are," and so on.	A
Choose (page 48)	When B has the words "for all," "for each," and so on.	A, and choose an object with the certain property.
Specialization (page 59)	When A has the words "for all," "for each," and so on.	A

What to Conclude	How to Do It
B	Work forward from A and apply the backward process to B.
NOT A	Work forward from *NOT B* and backward from *NOT A*.
Some contradiction	Work forward from A and *NOT B* to reach a contradiction.
That there is the desired object	Guess, construct, and so on, the object. Then show that it has the certain property and that the something happens.
That the something happens	Work forward from A and the fact that the object has the certain property. Also work backward from the something that happens.
B	Work forward by specializing A to one particular object having the certain property.

Table 13.2 Summary of Special Proof Techniques

Proof Technique	When to Use It	What to Assume
Induction (page 107)	When a statement $P(n)$ is true for each integer n beginning with n_0.	$P(n)$ is true for n.
Direct Uniqueness (page 105)	When B has the word "unique" in it.	There are two such objects, and A
Indirect Uniqueness (see 108)	When B has the word "unique" in it.	There are two different objects, and A
Proof by Elimination (page 121)	When B has the form "C *OR* D."	A and *NOT* C or A and *NOT* D
Proof by Cases (page 122)	When A has the form "C *OR* D."	Case 1: C Case 2: D
Max/Min 1 (page 126)	When B has the form "$\max S \leq x$" or "$\min S \geq x$".	Choose an $s \in S$, and A
Max/Min 2 (page 126)	When B has the form "$\max S \geq x$" or "$\min S \leq x$".	A

What to Conclude	How to Do It
That $P(n+1)$ is true; also show that $P(n_0)$ is true	First show that $P(n_0)$ true. Then use the assumption that $P(n)$ is true to prove that $P(n+1)$ is true.
That the two objects are equal	Work forward using A and the properties of the objects. Also work backward to show the objects are equal.
Some contradiction	Work forward from A using the properties of the two objects and the fact that they are different.
D	Work forward from A and *NOT* C, and backward from D.
or	or
C	Work forward from A and *NOT* D, and backward from C.
B B	First prove that C implies B; then prove that D implies B.
$s \leq x$ or $s \geq x$	Work forward from A and the fact that $s \in S$. Also work backward.
That there is an $s \in S$ for which $s \geq x$ or $s \leq x$	Use A and the construction method to produce the desired $s \in S$.

Exercises

Note: Solutions to exercises marked with a B are in the back of this book. Solutions to exercises marked with a W are located on the World Wide Web at http://www.wiley.com/college/solow/.

Note: All proofs should contain an analysis of proof and a condensed version. Definitions for all mathematical terms are provided in the glossary at the end of the book.

^{B}**13.1**   For each of the following statements, indicate which technique you would use to begin the proof and explain why.

   a. If p and q are odd integers, then the equation $x^2 + 2px + 2q = 0$ has no rational solution for x.

   b. For every integer $n \geq 4$, $n! > n^2$.

   c. If f and g are convex functions, then $f + g$ is a convex function.

   d. If a, b, and c are real variables, then the maximum value of $ab + bc + ac$ subject to the condition that $a^2 + b^2 + c^2 = 1$ is less than or equal to 1.

   e. In a plane, there is one and only one line perpendicular to a given line l through a point P on the line.

   f. If f and g are two functions such that (1) for all real numbers x, $f(x) \leq g(x)$ and (2) there is no real number M such that, for all x, $f(x) \leq M$, then there is no real number $M > 0$ such that, for all real numbers x, $g(x) \leq M$.

   g. If f and g are continuous functions at the point x, then so is the function $f + g$.

   h. If f and g are continuous functions at the point x, then for every real number $\epsilon > 0$, there is a real number $\delta > 0$, such that, for all real numbers y with $|x - y| < \delta$, $|f(x) + g(x) - (f(y) + g(y))| < \epsilon$.

   i. If f is the function defined by $f(x) = 2^x + \frac{x}{2}$, then there is a real number x^* between 0 and 1 such that, for all y, $f(x^*) \leq f(y)$.

13.2   For each of the problems in Exercise 13.1, state how the technique you chose to begin the proof would be applied to that problem. That is, indicate what you would assume, what you would conclude, and how you go about doing it.

^{W}**13.3**   Describe how you would use each of the following techniques to prove that "for every integer $n \geq 4$, $n! > n^2$." State what you would assume and what you would conclude.

a. Induction method.

b. Choose method.

c. Forward-Backward method. (Hint: Convert the problem to an equivalent one of the form, "if ... then ...".)

d. Contradiction method.

13.4   Suppose the forward-backward method is used to start each of the following proofs. List all of the techniques that are likely to be used subsequently in the proof.

a. (C AND D) implies (E OR F).

b. (C OR D) implies (E AND F).

c. If X is an object such that for all objects Y with a certain property, something happens, then there is an object Z with a certain property such that something else happens.

d. If for all objects X with a certain property, something happens, then there is an object Y with a certain property such that, for all objects Z with a certain property, something else happens.

13.5   Repeat the previous exercise assuming that the contrapositive method is used to start each proof.

Appendix A
Putting It All Together:
Part I

This appendix provides examples of how to read and understand written proofs as they might appear in a textbook or other mathematical literature. Some general strategies for creating your own proofs are also discussed and demonstrated.

A.1 HOW TO READ A PROOF

Due to the way in which they are written, there are three reasons why proofs are challenging to read:

1. The author does not always refer to the techniques by name.

2. Several steps are combined in a single sentence with little or no justification.

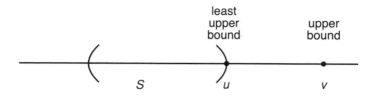

Fig. A.1   The least upper bound for a set.

3. The steps of a proof are not necessarily presented in the order in which they were performed when the proof was done.

To read a proof, you have to reconstruct the author's thought processes. Doing so requires that you identify which techniques are used and how they apply to the particular problem. The next example demonstrates how to read a proof. The proposition deals with the concept of the *least upper bound* of a set of real numbers, meaning the smallest value that is greater than or equal to all elements of the set (see Figure A.1). Formal definitions follow.

Definition 18 *A real number u is an* **upper bound** *for a set of real numbers T if and only if for all elements $t \in T$, $t \le u$.*

Definition 19 *A real number u is a* **least upper bound** *for a set S of real numbers if and only if (1) u is an upper bound for S and (2) for every upper bound v for S, $u \le v$.*

Proposition 25 *If u is an upper bound for a set S of real numbers and for every real number $\epsilon > 0$, there is an element $x \in S$ such that $x > u - \epsilon$, then u is a least upper bound for S.*

Proof of Proposition 25.   (For reference purposes, each sentence of the proof is written on a separate line.)

S1:   Suppose, to the contrary, that u is not a least upper bound for S.

S2:   So, let t be an upper bound for S with $t < u$.

S3:   Let $\epsilon = u - t > 0$.

S4:   Then, from the hypothesis, there is an element $x \in S$ such that $x > u - \epsilon = t$.

S5:   But then t is not an upper bound for S.

This contradiction completes the proof. □

Analysis of Proof. An interpretation of statements $S1$ through $S5$ follows.

Interpretation of S1: *Suppose, to the contrary, that u is not a least upper bound for S.*

The fact that the author says, "Suppose, to the contrary, ..." indicates that either the contradiction or contrapositive method is used. Reading ahead to the end of the proof, you can see that the contradiction method is used. At this point, you should identify the contradiction and keep it in mind as you read the rest of the proof. In this case, the contradiction is that the real number t is an upper bound for S (see $S2$) and yet t is not an upper bound for S (see $S5$). According to the contradiction method, the author should work forward from A and $NOT\ B$ to reach this contradiction. Indeed that is what is done in statements $S2$ through $S5$.

Interpretation of S2: *So, let t be an upper bound for S with $t < u$.*

The author is working forward from $NOT\ B$, that is, from the assumption that u is not a least upper bound for S. Using the techniques in Chapter 10 for finding the NOT of a statement containing the word AND, the NOT of Definition 19 for a least upper bound is the following:

> **A1:** Either u is not an upper bound for S or else there is an upper bound v for S with $v < u$.

Recognizing the key words "either/or" in the forward statement $A1$, you might consider proceeding by cases (see Chapter 12). Accordingly, you would assume first that u is not an upper bound for S. However, as stated in the hypothesis, u *is* an upper bound for S. Therefore this case cannot happen and so only the remaining case needs to be addressed. Thus, you can assume that there is an upper bound v for S with $v < u$. Indeed, this is what the author does by saying in $S2$ that

> **A2:** t is an upper bound for S with $t < u$.

Interpretation of S3: *Let $\epsilon = u - t > 0$.*

The author introduces the real number ϵ which is defined to be $u - t$. The author also notes that this value is greater than 0, which is true because $t < u$ (see $A2$). Although it is not clear at this point why and how this value of ϵ is used, keep in mind from $S5$ that the author is trying to reach the contradiction that

> **B1:** t is not an upper bound for S.

Interpretation of S4: *Then, from the hypothesis, there is an element $x \in S$ such that $x > u - \epsilon = t$.*

The author is now working forward from the for-all statement in the hypothesis. That is, the author is specializing that for-all statement to one particular value of $\epsilon > 0$, namely, the value of $\epsilon = u - t$ from $S3$. The result of specialization, as stated by the author in $S4$, is

A3: There is an $x \in S$ such that $x > u - \epsilon = u - (u - t) = t$.

Interpretation of S5: *But then t is not an upper bound for S.*

It is here that the author claims that t is not an upper bound for S. It remains to determine how the author is justified in making this statement. The reason is that the author has worked backward from $B1$ by writing the NOT of Definition 18 of an upper bound for a set. In particular, to show that t is not an upper bound for S, the author must show that

B2: There is an element $x \in S$ such that $x > t$.

Indeed, this element x is the one created in $A3$. In other words, recognizing the key words "there is" in $B2$, the author uses the construction method to produce the desired value for x. The author has therefore correctly reached the conclusion that t is not an upper bound for S, contradicting the fact that t is an upper bound for S (see $S2$), thus completing the proof.

Read the condensed proof again. Observe that what makes it challenging to understand is the lack of reference to the specific techniques that are used, especially the backward process and the specialization method. The author also omitted part of the thought process by combining several steps into a single sentence. To understand the proof you have to fill in the missing details.

In summary, when reading proofs, expect to do some work on your own. Learn to identify the various techniques that are used. Begin by trying to determine if the forward-backward or contradiction method is the primary technique. Then try to follow the methodology associated with that technique. Be watchful for quantifiers to appear, for then the corresponding choose, induction, construction, and/or specialization methods are likely to be used. The inability to follow a particular step of a written proof is often due to the lack of sufficient detail. To fill in the gaps, learn to ask yourself how you would proceed to do the proof. Then try to see if the written proof matches your thought process.

A.2   HOW TO DO A PROOF

Now that you know how to read a proof, it is time to see how to do a proof on your own. In the example that follows, pay particular attention to how the form of the statement under consideration dictates the technique to be used.

Definition 20 *A set S of real numbers is* **bounded** *if and only if there is a real number $M > 0$ such that, for all elements $x \in S$, $|x| < M$.*

Proposition 26 *If T is a bounded subset of real numbers, then any subset of T is bounded.*

Analysis of Proof. Looking at the statement B, you should recognize the quantifier "for any" and thus begin with the choose method. Accordingly, you would write, "Let S be a subset of T. It will be shown that S is bounded." In other words, you should choose

A1: A subset S of T,

for which it must be shown that

B1: S is bounded.

Not recognizing any key words in $B1$, it is probably best to proceed with the forward-backward method. The idea is to work backward from $B1$ until it is no longer fruitful to do so. Then apply the forward process to the hypothesis that T is a bounded set of real numbers.

Working backward from $B1$, can you pose a key question? One such question is, "How can I show that a set (namely, S) is bounded?" Using Definition 20 means you must show that

B2: There is a real number $M > 0$ such that, for all elements $x \in S$, $|x| < M$.

The appearance of the first quantifier "there is" in $B2$ suggests using the construction method and so you should turn to the forward process to produce the desired value for M. Working forward from the hypothesis that T is bounded, by Definition 20, you can state that

A2: There is a real number $M' > 0$ such that, for all elements $x \in T$, $|x| < M'$.

Observe that the symbol M' is used instead of M because M is already used in $B2$.

Perhaps M' is the desired value of M. When using the construction method, it is advisable to construct the desired object in an "obvious" way, if possible. In this case, that means setting

A3: $M = M'$.

If this guess is correct, then you must still show that the certain property in $B2$ holds (that is, $M > 0$) and also that the something happens (that is, that for all $x \in S$, $|x| < M$). It is not hard to see that $M > 0$ because $M' > 0$ and $M = M'$. Thus it remains only to show that

B3: For all elements $x \in S$, $|x| < M = M'$.

The appearance of the quantifier "for all" in $B3$ suggests proceeding with the choose method, whereby you would choose

A4: An element $x' \in S$,

for which you must show that

B4: $|x'| < M$, or, because $M = M'$, $|x'| < M'$.

Observe that the symbol x' is used for the chosen object to avoid confusion with the general symbol x. At this point it might not be clear how to reach the conclusion that $|x'| < M'$, but look carefully at the for-all statement in $A2$. You can use specialization to reach the desired conclusion that $|x'| < M'$, provided that you can specialize the statement to $x = x'$. To be able to do so, you must make sure that the particular object under consideration (namely, x') does satisfy the certain property in the for-all statement (namely, that $x' \in T$). Your objective, therefore, is to show that

B5: $x' \in T$.

Recall that x' is chosen to be an element of S (see $A4$) and that S is chosen to be a subset of T (see $A1$). How can you use this information to show that $x' \in T$? Working forward from the definition for a subset, you know that

A5: For all $x \in S$, $x \in T$.

Thus, you can specialize $A5$ to $x = x'$, which you know is in S (see $A4$), and reach the desired conclusion that $x' \in T$, completing the proof.

Proof of Proposition 26.   Let S be a subset of T. To show that S is bounded, a real number $M > 0$ is produced with the property that, for all $x \in S$, $|x| < M$.

By hypothesis, T is bounded, and so there is a real number $M' > 0$ such that, for all $x \in T$, $|x| < M'$. Set $M = M'$. To see that M has the desired properties, let x' be an element of S. Because S is a subset of T, $x' \in T$. But then it follows that $|x'| < M' = M$, as desired. ☐

A.3   GENERAL GUIDELINES FOR CREATING A PROOF

Creating proofs is not a precise science. Some general suggestions are provided here. When trying to prove that "A implies B," consciously choose a technique based on the form of B. If no key words are apparent, then it is probably best to proceed with the forward-backward method. If unsuccessful, there are several avenues to pursue before giving up. You might try asking yourself why B cannot be false, thus leading you to the contradiction (or contrapositive) method.

The point is that, when you are unable to complete a proof for one reason or another, it is advisable to seek another technique consciously. Sometimes you can manipulate the form of the statement B so as to induce a different technique. For instance, suppose that the statement B contains the quantifier "for all" in the standard form:

B : For all "objects" with a "certain property,"
"something happens."

If the choose method fails, then you might try rewriting B so that it reads:

B : There is no "object" with the "certain property" such that
the "something" does not happen.

Now the contradiction method suggests itself because the key word "no" appears. Thus you would assume that there is an object with the certain property such that the something does not happen. Your objective is to reach a contradiction.

When the statement B has recognizable key words, it is best to try the corresponding technique first. Remember that, as you proceed through a proof, different techniques are needed as the form of the statement under consideration changes.

If you are really stuck, it can sometimes be advantageous to leave the problem for a while, for when you return, you might see a new approach. Undoubtedly you will learn many tricks of your own as you solve more and more problems.

Appendix B
Putting It All Together:
Part II

This appendix provides additional practice in reading and doing proofs. The examples presented here are more difficult than those in Appendix A. The additional difficulty is due, in part, to the fact that the statements under consideration contain three nested quantifiers as opposed to two.

B.1 HOW TO READ A PROOF

The first example is designed to teach you how to read a condensed proof as it might appear in a textbook. The example deals with the concept of a *continuous* function of one variable, meaning that if the value of the variable is changed slightly, then the value of the function does not change radically. Another informal way of saying that a function is continuous is that you can draw the graph of the function without lifting the pencil. To proceed formally, a mathematical definition of continuity is needed.

Definition 21 *A function f of one variable is* **continuous** *at the point x if and only if for every real number $\epsilon > 0$, there is a real number $\delta > 0$ such that, for all real numbers y with the property that $|x - y| < \delta$, it follows that $|f(x) - f(y)| < \epsilon$.*

Proposition 27 *If f and g are two functions that are continuous at x, then the function $f + g$ is also continuous at x. [Note: $f + g$ is the function whose value at any point y is $f(y) + g(y)$.]*

Proof of Proposition 27. (For reference purposes, each sentence of the proof is written on a separate line.)

S1:   To see that the function $f + g$ is continuous at x, let $\epsilon > 0$.

S2:   It is shown that there is a $\delta > 0$ such that, for all y with $|x - y| < \delta$, $|f(x) + g(x) - [f(y) + g(y)]| < \epsilon$.

S3:   Because f is continuous at x, there is a $\delta_1 > 0$ such that, for all y with $|x - \overset{.}{y}| < \delta_1$, $|f(x) - f(y)| < \epsilon/2$.

S4:   Similarly, because g is continuous at x, there is a $\delta_2 > 0$ such that, for all y with $|x - y| < \delta_2$, $|g(x) - g(y)| < \epsilon/2$.

S5:   Let $\delta = \min\{\delta_1, \delta_2\} > 0$, and let y have the property that $|x - y| < \delta$.

S6:   Then $|x - y| < \delta_1$ and $|x - y| < \delta_2$, and so

$$
\begin{aligned}
|f(x) + g(x) - [f(y) + g(y)]| \;\; &\leq \;\; |f(x) - f(y)| + \\
&\quad\; |g(x) - g(y)| \\
&< \;\; \epsilon/2 + \epsilon/2 \\
&= \;\; \epsilon.
\end{aligned}
$$

The proof is now complete. $\Box$

Analysis of Proof.   An interpretation of each of the statements $S1$ through $S6$ follows.

Interpretation of S1. *To see that the function $f + g$ is continuous at x, let $\epsilon > 0$.*

This statement indicates that the forward-backward method is used because the statement, "to see that $f + g$ is continuous at x," means that the author is about to show that the statement B is true. The question is why did the author say, "... let $\epsilon > 0$"? Hopefully it is related to showing that $f + g$ is continuous at x. What has happened is that the author has implicitly posed the key question, "How can I show that a function (namely, $f + g$) is continuous at a point (namely, x)?" and has used Definition 21 to provide the answer, thus it must be shown that

B1: For every $\epsilon > 0$, "something happens."

Recognizing the appearance of the quantifier "for every" in the backward process, you should then proceed with the choose method, whereby you choose

A1: A real number $\epsilon > 0$.

The author has done this by saying "... let $\epsilon > 0$." Observe that the symbol ϵ appearing in the definition of continuity refers to a general value for ϵ, while the symbol ϵ appearing in $S1$ refers to a specific value resulting from the choose method. The double use of a symbol can be confusing but is common in written proofs.

Interpretation of S2. *It is shown that there is a $\delta > 0$ such that, for all y with $|x - y| < \delta$, $|f(x) + g(x) - [f(y) + g(y)]| < \epsilon$.*
The author is stating precisely what is going to be done, but why should the author want to show that "there is a $\delta > 0$ such that ..."? Recall that, in $S1$, the choose method is used to select an $\epsilon > 0$ (see $A1$). Accordingly, it must be shown that, for that chosen ϵ, the something happens. From Definition 21, that something is

B2: There is a $\delta > 0$ such that, for all y with $|x - y| < \delta$,
$|f(x) + g(x) - [f(y) + g(y)]| < \epsilon$.

Interpretation of S3. *Because f is continuous at x, there is a $\delta_1 > 0$ such that, for all y with $|x - y| < \delta_1$, $|f(x) - f(y)| < \epsilon/2$.*
The author has turned to the forward process (having recognized from $B2$ that the quantifier "there is" suggests using the construction method to produce the desired δ). The author is working forward from the hypothesis that f is continuous at x using Definition 21. What the author has failed to say is that, in the definition, the quantifier "for all" arises and the statement is specialized to the value of $\epsilon/2$, where ϵ is the one chosen in $A1$, thus leading to

A2: There is a $\delta_1 > 0$ such that, for all y with $|x - y| < \delta_1$,
$|f(x) - f(y)| < \epsilon/2$.

At this moment, it is not clear why the author specializes to the value of $\epsilon/2$ as opposed to ϵ. The reason becomes clear in $S6$.

Interpretation of S4. *Similarly, because g is continuous at x, there is a $\delta_2 > 0$ such that, for all y with $|x - y| < \delta_2$, $|g(x) - g(y)| < \epsilon/2$.*

The author is working forward from the fact that g is continuous at x in the same way that is done in $S3$. Thus the author can conclude that

A3: There is a $\delta_2 > 0$ such that, for all y with $|x - y| < \delta_2$, $|g(x) - g(y)| < \epsilon/2$.

Keep in mind that, from $B2$, the author is trying to construct a value of $\delta > 0$ for which something happens.

Interpretation of S5. *Let $\delta = \min\{\delta_1, \delta_2\} > 0$, and let y have the property that $|x - y| < \delta$.*
It is here that the desired value for $\delta > 0$ is produced. Specifically, the values of δ_1 and δ_2 from $A2$ and $A3$, respectively, are combined to give the desired δ, namely,

A4: $\delta = \min\{\delta_1, \delta_2\}$.

It is not yet clear why δ is constructed in this way. In any event, it is the author's responsibility to show that $\delta > 0$ and also that δ satisfies the something that happens associated with the quantifier "there is" in $B2$, that is,

B3: For all y with $|x - y| < \delta$, $|f(x) + g(x) - [f(y) + g(y)]| < \epsilon$.

In $S5$ the author notes that $\delta > 0$ (which is true because both $\delta_1, \delta_2 > 0$). Also, the reason that the author says, "... let y have the property that $|x - y| < \delta$" is because the choose method is used to show that $B3$ is true. Thus, the author has chosen

A5: A real number y with $|x - y| < \delta$.

Note again the double use of the symbol y. Nonetheless, to complete the proof, it must be shown that

B4: $|f(x) + g(x) - [f(y) + g(y)]| < \epsilon$.

Interpretation of S6. *Then $|x - y| < \delta_1$ and $|x - y| < \delta_2$, and so*

$$
\begin{aligned}
|f(x) + g(x) - [f(y) + g(y)]| \;&\leq\; |f(x) - f(y)| + \\
&\quad\; |g(x) - g(y)| \\
&<\; \epsilon/2 + \epsilon/2 \\
&=\; \epsilon.
\end{aligned}
$$

It is here that the author concludes that $B4$ is true. The only question is how? It is not hard to see that

A6: $|f(x) + g(x) - [f(y) + g(y)]| \leq |f(x) - f(y)| + |g(x) - g(y)|$

because the absolute value of the sum of the two numbers $f(x) - f(y)$ and $g(x) - g(y)$ is less than or equal to the sum of the absolute values of those two numbers. But why is $|f(x) - f(y)| + |g(x) - g(y)| < \epsilon/2 + \epsilon/2$? Once again the author has omitted part of the thought process.

What has happened is that the author uses specialization to claim that $|f(x) - f(y)| < \epsilon/2$ and $|g(x) - g(y)| < \epsilon/2$. Specifically, the for-all statements in $A2$ and $A3$ are both specialized to the particular value of y chosen in $A5$. Recall that, when using specialization, one must be sure that the particular object under consideration (namely, y) satisfies the certain property (in this case, $|x - y| < \delta_1$ and $|x - y| < \delta_2$). Note that, in $S6$, the author does claim that $|x - y| < \delta_1$ and $|x - y| < \delta_2$, but can you see why this is true? Look at $A4$ and $A5$. Because y is chosen so that $|x - y| < \delta$ and $\delta = \min\{\delta_1, \delta_2\}$, it must be that both $\delta \leq \delta_1$ and $\delta \leq \delta_2$, so

A7: $|x - y| < \delta_1$ and $|x - y| < \delta_2$.

Hence, the author is justified in specializing the for-all statements in $A2$ and $A3$ to this particular value of y and hence can conclude that

A8: $|f(x) - f(y)| < \epsilon/2$ and
A9: $|g(x) - g(y)| < \epsilon/2$.

In other words, the author is justified in claiming that

$$\begin{aligned}
|f(x) + g(x) \quad [f(y) + g(y)]| &\leq |f(x) - f(y)| + |g(x) - g(y)| \\
&< \epsilon/2 + \epsilon/2 \\
&= \epsilon.
\end{aligned}$$

Perhaps now it is clear why the author specialized the definitions of continuity for f and g to $\epsilon/2$ instead of ϵ (see $A2$ and $A3$).

Again, note that what makes $S6$ so challenging to understand is the omission of part of the thought process. You must learn to fill in the missing links by determining which techniques are used and consequently what has to be done.

B.2   HOW TO DO A PROOF

The next example illustrates how to go about creating your own proof. Pay particular attention to how the form of the statement under consideration leads to the correct technique.

Proposition 28 *If f is a function that, at the point x, satisfies the property that there are real numbers $c > 0$ and $\delta' > 0$ such that, for all y with $|x - y| < \delta'$, $|f(x) - f(y)| \leq c|x - y|$, then f is continuous at x.*

Analysis of Proof.   Begin by looking at the statement B and trying to select a technique. Because B does not have a particular form, use the forward-backward method. Thus you are led to the key question, "How can I show that a function (namely, f) is continuous at a point (namely, x)?" Definition 21 provides the answer, so you must show that

> **B1:** For all $\epsilon > 0$, there is a $\delta > 0$ such that, for all y with $|x - y| < \delta$, $|f(x) - f(y)| < \epsilon$.

Looking at $B1$, you should recognize the "for all" as the first of the nested quantifiers. Hence you should use the choose method to choose

> **A1:** A real number $\epsilon' > 0$.

It is advisable to use a symbol other than ϵ for the specific choice so as to avoid confusion. In the condensed proof you would write, "Let $\epsilon' > 0 \ldots$" Once you have selected $\epsilon' > 0$, the choose method requires you to show that the something happens, in this case that

> **B2:** There is a $\delta > 0$ such that, for all y with $|x - y| < \delta$, $|f(x) - f(y)| < \epsilon'$.

Looking at $B2$, you should recognize the key words "there is" as the next nested quantifier. Accordingly, use the construction method to produce the desired value of δ. When the construction method is used, it is advisable to turn to the forward process to produce the desired object.

Working forward from the hypothesis that there are real numbers $c > 0$ and $\delta' > 0$ for which something happens, you should attempt to construct δ. Perhaps $\delta = \delta'$ (or perhaps $\delta = c$). Why not guess that

> **A2:** $\delta = \delta'$.

If this guess is not correct, then maybe you will discover what the proper value for δ should be. Alternatively, you could attempt to construct δ by working backward from the fact that you want δ to satisfy the property that, for all y with $|x - y| < \delta$, $|f(x) - f(y)| < \epsilon'$.

In any event, to check if the current guess of $\delta = \delta'$ is correct, it is necessary to see if the certain property associated with the quantifier "there is" in $B2$ holds for δ' and also if the something happens. From $B2$, the certain property is that $\delta > 0$, but because $\delta = \delta'$ and $\delta' > 0$, it must be that $\delta > 0$. It remains only to verify that, for $\delta = \delta'$,

> **B3:** For all y with $|x - y| < \delta$, $|f(x) - f(y)| < \epsilon'$.

Recognizing the quantifier "for all" in the backward process in $B3$, use the choose method to select

> **A3:** A real number y' with $|x - y'| < \delta$,

for which it must be shown that

B4: $|f(x) - f(y')| < \epsilon'$.

To see that $B4$ is true, the hypothesis still has some unused information. Specifically, the number $c > 0$ has not been used, nor has the statement:

A4: For all y with $|x - y| < \delta'$, $|f(x) - f(y)| \leq c|x - y|$.

On recognizing the quantifier "for all" in the forward process in $A4$, consider using specialization—the only question is which value of y to specialize to. Why not try the value of y' chosen in $A3$? Fortunately, y' was chosen to have the certain property in $A4$ (that is, $|x - y'| < \delta$ and $\delta = \delta'$, so $|x - y'| < \delta'$), and so you can specialize $A4$ to y' obtaining:

A5: $|f(x) - f(y')| \leq c|x - y'|$.

Recall that the last statement obtained in the backward process is $B4$, so the idea is to see if you can make $A5$ look more like $B4$. For instance, because y' is chosen in $A3$ so that $|x - y'| < \delta$, you can rewrite $A5$ as

A6: $|f(x) - f(y')| < c\delta$.

You can obtain $B4$ if $c\delta$ were known to be $\leq \epsilon'$. While from $A2$ you know that $\delta = \delta'$, you do not know that $c\delta' \leq \epsilon'$ This means that the original guess of $\delta = \delta'$ in $A2$ is incorrect. Perhaps δ should be chosen > 0 and with the property that $c\delta \leq \epsilon'$, or equivalently, because $c > 0$, why not construct

A2: $0 < \delta \leq \epsilon'/c$.

To see if the new guess of $0 < \delta \leq \epsilon'/c$ is correct, it is necessary to check if, for this value of δ, the certain property associated with the quantifier "there is" in $B2$ holds and also whether the something happens. Clearly δ is chosen to be > 0 and thus has the certain property. It remains only to verify that the something happens, in this case, that $B3$ is true.

As before, one would proceed with the choose method to select y' with $|x - y'| < \delta$ (see $A3$), and again one is led to showing that $B4$ is true. The previous approach was to specialize $A4$ to y'. This time, however, there is a problem because y' is not known to satisfy the certain property in $A4$, that $|x - y'| < \delta'$. Unfortunately, all you do know is that $|x - y'| < \delta$ (see $A3$) and $0 < \delta \leq \epsilon'/c$ (see the "new" $A2$). If only you could apply specialization to $A4$, then you would obtain $A5$, and hence $A6$. Finally, because $c\delta \leq \epsilon'$, you could reach the desired conclusion that $B4$ is true.

Can you figure out how to choose δ so that both $c\delta \leq \epsilon'$ and you can specialize $A1$ to y'? The answer is to construct

A2: $0 < \delta \leq \min\{\delta', \epsilon'/c\}$,

for then, when y' is chosen with $|x - y'| < \delta$ in $A3$, it follows that $|x - y'| < \delta'$ because $\delta \leq \delta'$. Thus, you can specialize $A4$ to y'. Also, from $A5$ and the fact that $\delta \leq \epsilon'/c$, you can conclude that $|f(x) - f(y')| \leq c|x - y'| < c\delta \leq c(\epsilon'/c) = \epsilon'$, thus completing the proof.

In the condensed proof that follows, observe that the construction of the correct value for δ is presented at the beginning of the proof—in the second sentence—which does not correspond to the order in which this value is constructed in the foregoing analysis.

Proof of Proposition 28.   To show that f is continuous at x, let $\epsilon' > 0$. By the hypothesis that $c > 0$, it is possible to construct $0 < \delta \leq \min\{\delta', \epsilon'/c\}$. Then, for y' with $|x - y'| < \delta$, it follows that $|x - y'| < \delta'$ and so, from the hypothesis, $|f(x) - f(y')| \leq c|x - y'| < c\delta$. Furthermore, because $\delta \leq \epsilon'/c$, $|f(x) - f(y')| < c\delta \leq c(\epsilon'/c) = \epsilon'$, and the proof is complete. $\square$

As with any language, reading, writing, and speaking come only with practice. The techniques presented in this book are designed to get you started in the right direction.

Solutions to Selected Exercises

Solutions to Selected Exercises in Chapter 1

1.1 (a), (c), (e), and (f) are statements.

1.3 a. Hypothesis: The right triangle XYZ with sides of lengths x and y and hypotenuse of length z has an area of $z^2/4$.

Conclusion: The triangle XYZ is isosceles.

b. Hypothesis: n is an even integer.
Conclusion: n^2 is an even integer.

c. Hypothesis: a, b, c, d, e, and f are real numbers for which $ad - bc \neq 0$.

Conclusion: The two linear equations $ax + by = e$ and $cx + dy = f$ can be solved for x and y.

1.6 a. True because A $(2 > 7)$ is false.

b. True because A $(2 < 7)$ and B $(1 < 3)$ are both true.

c. True because B $(1 < 2)$ is true (the truth of A does not matter).

d. True if $x \neq 3$ because then A $(x = 3)$ is false.
False if $x = 3$ because then A $(x = 3)$ is true and B is false.

1.8 If you want to prove that "A implies B" is true and you know that B is false, then A should also be false. The reason is that if A is false, then it does not matter whether B is true or false because Table 1.1 ensures that "A implies B" is true. On the other hand, if A is true and B is false, then "A implies B" would be false.

1.10   (T = true, F = false)

A	B	C	$(B \Rightarrow C)$	$A \Rightarrow (B \Rightarrow C)$
T	T	T	T	T
T	T	F	F	F
T	F	T	T	T
T	F	F	T	T
F	T	T	T	T
F	T	F	F	T
F	F	T	T	T
F	F	F	T	T

Solutions to Selected Exercises in Chapter 2

2.3 (c) is incorrect because it uses the specific notation given in the problem.

2.5 (a) is correct because it asks, without symbols, how to prove that the statement B is true. Questions (b) and (c) are not valid because they use the specific notation in the problem. Question (d) is an incorrect question for this problem.

2.7 a. How can I show that two lines are parallel?
How can I show that two lines do not intersect?
How can I show that two lines tangent to a circle are parallel?
How can I show that two tangent lines passing through the endpoints of the diameter of a circle are parallel?

b. How can I show that a function is continuous?
How can I show that the sum of two functions is continuous?
How can I show that the sum of two continuous functions is continuous?

2.10 a. Show that the difference of the two numbers equals zero.
Show that one number is $\leq$ the other number and vice versa.
Show that the ratio of the two numbers is 1.
Show that the two numbers are both equal to a third number.

b. Show that their corresponding side-angle-sides are equal.
Show that their corresponding angle-side-angles are equal.
Show that their corresponding side-side-sides are equal.
Show that they are both congruent to a third triangle.

2.15 a. $(x - 2)(x - 1) < 0$.
$x(x - 3) < -2$.
$-x^2 + 3x - 2 > 0$.

b. $x/z = 1/\sqrt{2}$.
Angle X is a 45-degree angle.
$\cos(X) = 1/\sqrt{2}$.

c. The circle has its center at (3,2).
The circle has a radius of 5.
The circle crosses the y-axis at (0,6) and (0, -2).
$x^2 - 6x + 9 + y^2 - 4y + 4 = 25$.

2.17 (d) is not valid because "$x \neq 5$" is not stated in the hypothesis, and so, if $x = 5$, it will not be possible to divide by $x - 5$.

2.19 a. **Analysis of Proof.** A key question associated with the conclusion is, "How can I show that a real number (namely, x) is 0?" To show that $x = 0$, it will be established that

$B1: x \leq 0$ and $x \geq 0$.

Working forward from the hypothesis immediately establishes that

$A1: x \geq 0$.

To see that $x \leq 0$, it will be shown that

$B2: x = -y$ and $-y \leq 0$.

Both of these statements follow by working forward from the hypotheses that

$A2: x + y = 0$ (so $x = -y$) and
$A3: y \geq 0$ (so $-y \leq 0$).

It remains only to show that

$B3: y = 0$,

which follows by working forward from the fact that

$A4: x = 0$

and the hypothesis that

$A5: x + y = 0$,

so,

$A6: 0 = x + y = 0 + y = y$.

b. **Proof.** To see that both $x = 0$ and $y = 0$, it will first be shown that $x \geq 0$ (which is given in the hypothesis) and $x \leq 0$. The latter is accomplished by showing that $x = -y$ and that $-y \leq 0$. To see that $x = -y$, observe that the hypothesis states that $x + y = 0$. Similarly, $-y \leq 0$ because the hypothesis states that $y \geq 0$. Thus, $x = 0$. To see that $y = 0$, one can substitute $x = 0$ in the hypothesis $x + y = 0$ to reach the desired conclusion. ◻

2.23 Analysis of Proof. In this problem, one has:

A : The right triangle XYZ is isosceles.
B : The area of triangle XYZ is $z^2/4$.

A key question for B is, "How can I show that the area of a triangle is equal to a particular value?" One answer is to use the formula for computing the area of a triangle to show that

$B1$: $z^2/4 = xy/2$.

Working forward from the hypothesis that triangle XYZ is isosceles, one has that

$A1$: $x = y$, so
$A2$: $x - y = 0$.

Because XYZ is a right triangle, from the Pythagorean theorem,

$A3$: $z^2 = x^2 + y^2$.

Squaring both sides of the equality in $A2$ and performing algebraic manipulations yields

$A4$: $(x - y)^2 = 0$.
$A5$: $x^2 - 2xy + y^2 = 0$.
$A6$: $x^2 + y^2 = 2xy$.

Substituting $A3$ in $A6$ yields

$A7$: $z^2 = 2xy$.

Dividing both sides by 4 finally yields the desired result:

$A8$: $z^2/4 = xy/2$.

Proof. From the hypothesis, $x = y$, or equivalently, $x - y = 0$. Performing algebraic manipulations yields $x^2 + y^2 = 2xy$. By the Pythagorean theorem, $z^2 = x^2 + y^2$ and on substituting z^2 for $x^2 + y^2$, one obtains $z^2 = 2xy$, or, $z^2/4 = xy/2$. From the formula for the area of a right triangle, the area of $XYZ = xy/2$. Hence $z^2/4$ is the area of the triangle.
◻

Solutions to Selected Exercises in Chapter 3

3.1 a. Key Qn: How can I show that an integer is odd?

Abs. Ans: Show that the integer equals two times some integer plus one.

Spec. Ans: Show that $n^2 = 2k + 1$ for some integer k.

b. Key Qn: How can I show that a real number is rational?

Abs. Ans: Show that the real number is equal to the ratio of two integers in which the denominator is not zero.

Spec. Ans: Show that $s/t = p/q$, where p and q are integers and $q \neq 0$.

c. Key Qn: How can I show that two pairs of real numbers are equal?

Abs. Ans: Show that the first and second elements of one pair of real numbers are equal to the corresponding elements of the other pair.

Spec. Ans: Show that $x_1 = x_2$ and $y_1 = y_2$.

d. Key Qn: How can I show that an integer is prime?

Abs. Ans: Show that the integer is greater than 1 and can be divided only by itself and 1.

Spec. Ans: Show that $n > 1$ and, if k is an integer that divides n, then $k = 1$ or $k = n$.

e. Key Qn: How can I show that an integer divides another integer?

Abs. Ans: Show that the second integer equals the product of the first integer with another integer.

Spec. Ans: Show that the following expression is true for some integer $k : (n-1)^3 + n^3 + (n+1)^3 = 9k$.

3.5 (T = true, F = false)

a. Truth Table for the Converse of "A Implies B."

A	B	"A Implies B"	"B Implies A"
T	T	T	T
T	F	F	T
F	T	T	F
F	F	T	T

b. Truth Table for the Inverse of "A Implies B."

A	B	$NOT\ A$	$NOT\ B$	"$NOT\ A$ Implies $NOT\ B$"
T	T	F	F	T
T	F	F	T	T
F	T	T	F	F
F	F	T	T	T

The converse and inverse of "A implies B" are equivalent. Both are true except when A is false and B is true.

3.7 a. If n is an odd integer, then n^2 is odd.

b. If r is a real number such that $r^2 \neq 2$, then r is rational.

c. If the quadrilateral $ABCD$ is a rectangle, then $ABCD$ is a parallel-ogram with one right angle.

3.10 Analysis of Proof. The forward-backward method gives rise to the key question, "How can I show that the square of an integer [namely, $(a + b)$] is odd?" One answer is to use the proposition in Exercise 3.9 whose conclusion is that n^2 is odd. Matching up the notation of $n = a + b$, you must show that the hypothesis of the proposition in Exercise 3.9 is true, that is, you must show that

$B1 : a + b$ is odd.

Turning to the forward process, from the hypothesis you know that a and b are consecutive integers, so

$A1 :$ One of a and b is even and the other is odd.

It therefore follows from $A1$ that $B1$ is true because an even plus an odd is odd, and so the proof is complete.

Proof. From the hypothesis that a and b are consecutive integers, it follows that one of a and b is even and the other is odd. Consequently, $a + b$ is odd, and so the hypothesis of the proposition in Exercise 3.9 is true. It follows that the conclusion of that proposition is also true, so $(a + b)^2$ is odd, and hence the proof is complete. $\square$

3.12 Analysis of Proof. Using the forward-backward method one is led to the key question, "How can I show that one statement (namely, A) implies another statement (namely, C)?" According to Table 1.1 the answer is to assume that the statement to the left of the word "implies" is true, and then reach the conclusion that the statement to the right of the word "implies" is true. In this case, you assume

$A1 : A$ is true,

and try to reach the conclusion that

$B1: C$ is true.

Working forward from the information in the hypothesis, because "A implies B" is true and A is true, by row 1 of Table 1.1 it must be that

$A1: B$ is true.

Because B is true, and "B implies C" is true, it must also be that

$A2: C$ is true.

Hence the proof is complete.

Proof. To conclude that "A implies C" is true, assume that A is true. By hypothesis, "A implies B" is true, so B must be true. Finally, because "B implies C" is true, it must be that C is true, thus completing the proof. $\square$

3.18 Analysis of Proof. The forward-backward method gives rise to the key question, "How can I show that a triangle is isosceles?" Using the definition of an isosceles triangle, you must show that two of its sides are equal, which, in this case, means you must show that

$B1: u - v.$

Working forward from the hypothesis, you know that

$A1: \sin(U) = \sqrt{u/2v}.$

By the definition of sine, $\sin(U) = u/w$, so

$A2: \sqrt{u/2v} = u/w,$

and by algebraic manipulations, one obtains that

$A3: w^2 = 2uv.$

Furthermore, by the Pythagorean theorem,

$A4: u^2 + v^2 = w^2.$

Substituting for w^2 from $A3$ in $A4$ one has that

$A5: u^2 + v^2 = 2uv$, or,
$A6: u^2 - 2uv + v^2 = 0.$

On factoring $A6$, and then taking the square root of both sides of the equality, it follows that

$A7: u - v = 0$

and so $u = v$, completing the proof.

Proof. Because $\sin(U) = \sqrt{u/2v}$ and also $\sin(U) = u/w$, $\sqrt{u/2v} = u/w$, or, $w^2 = 2uv$. Now from the Pythagorean theorem, $w^2 = u^2 + v^2$, and on substituting $2uv$ for w^2 and then performing algebraic manipulations, one has $u = v$. Thus the triangle is isosceles. $\square$

Solutions to Selected Exercises in Chapter 4

4.1

	Object	Certain Property	Something Happens
(a)	two people	none	they have the same number of friends
(b)	integer x	none	$f(x) = 0$
(c)	a point (x, y)	$x \geq 0$ and $y \geq 0$	$y = m_1 x + b_1$ and $y = m_2 x + b_2$
(d)	angle t'	$0 \leq t' \leq \pi$	$\tan(t') > \tan(t)$
(e)	integers m and n	none	$am + bn = c$

4.5 The construction method arises in the proof of Proposition 2 because the statement,

   $B1:$ n^2 can be expressed as two times some other integer

can be rewritten as follows to contain the quantifier "there is":

   $B1:$ There is an integer p such that $n^2 = 2p$.

The construction method is then used to produce the integer p for which $n^2 = 2p$. Specifically, using the fact that n is even and hence that there is an integer k such that $n = 2k$, the desired value for p is constructed as $p = 2k^2$. This value for p is correct because $n^2 = (2k)^2 = 4k^2 = 2(2k^2) = 2p$.

4.6 **Analysis of Proof.** The appearance of the key words "there is" in the conclusion suggests using the construction method to find an integer x for which $x^2 - 5x/2 + 3/2 = 0$. Factoring means that you want an integer x such that $(x - 1)(x - 3/2) = 0$. Thus, the desired value is $x = 1$. On substituting this value of x in $x^2 - 5x/2 + 3/2$ yields 0, so the quadratic equation is satisfied. This integer solution is unique because the only other solution is $x = 3/2$, which is not integer.

Proof. Factoring $x^2 - 5x/2 + 3/2$ means that the only roots are $x = 1$ and $x = 3/2$. Thus, there exists an integer, namely, $x = 1$, such that $x^2 - 5x/2 + 3/2 = 0$. The integer is unique. $\square$

4.9   Analysis of Proof. The forward-backward method gives rise to the key question, "How can I show that an integer (namely, a) divides another integer (namely, c)?" By the definition, one answer is to show that

$B1$: There is an integer k such that $c = ak$.

The appearance of the quantifier "there is" in $B1$ suggests turning to the forward process to construct the desired k.

From the hypothesis that $a|b$ and $b|c$, and by definition,

$A1$: There are integers p and q such that $b = ap$ and $c = bq$.

Therefore, it follows that

$A2$: $c = bq = (ap)q = a(pq)$,

and so the desired integer k is $k = pq$.

Proof. Because $a|b$ and $b|c$, by definition, there are integers p and q for which $b = ap$ and $c = bq$. But then it follows that $c = bq = (ap)q = a(pq)$, and so $a|c$. $\square$

4.13 The error in the proof occurs when the author says to divide by p^2 because it will not be possible to do so if $p = 0$, which can happen.

Solutions to Selected Exercises in Chapter 5

5.1 a. Object: real number x.
              Certain property: none.
              Something happens: $f(x) \leq f(x^*)$.
         b. Object: element x.
              Certain property: $x \in S$.
              Something happens: $g(x) \geq f(x)$.
         c. Object: element x.
              Certain property: $x \in S$.
              Something happens: $x \leq u$.
         d. Object: elements x and y, and real numbers t.
              Certain property: x and y in C, and $0 \leq t \leq 1$.
              Something happens: $tx + (1 - t)y$ is an element of C.
         e. Object: real numbers x, y, and t.
              Certain property: $0 \leq t \leq 1$.
              Something happens: $f(tx + (1 - t)y) \leq tf(x) + (1 - t)f(y)$.

5.3 a. Let x' be a real number. It will be shown that $f(x') \leq f(x^*)$.
         b. Let $x' \in S$. It will be shown that $g(x') \geq f(x')$.

c. Let $x' \in S$. It will be shown that $x' \leq u$.

d. Let x', $y' \in C$, and let t' be a real number with $0 \leq t' \leq 1$. It will be shown that $t'x' + (1 - t')y' \in C$.

e. Let x', y', t' be real numbers with $0 \leq t' \leq 1$. It will be shown that $f(t'x' + (1 - t')y') \leq t'f(x') + (1 - t')f(y')$.

5.5 When using the choose method to show that "for all objects with a certain property, something happens," you would choose one particular object that does have the certain property. You would then work forward from the certain property to reach the conclusion that the something happens. This is precisely the same as using the forward-backward method to show that "if X is an object with the certain property, then the something happens," whereby you would work forward from the fact that X is an object with the certain property and backward from the something that happens. In other words, you can convert a statement containing the quantifier "for all" to an equivalent statement having the form "if ... then ...". In the event that there is no object having the certain property in the for-all statement, then the hypothesis in the corresponding statement, "if X is an object with the certain property, then the something happens," is false. Thus, from the truth table, the "if ... then ..." statement is true and hence so is the original for-all statement.

5.8 The choose method is used in the first sentence of the proof where it says, "Let x be a real number." More specifically, the author asked the key question, "How can show that a real number (namely, x^*) is a maximum of a function (namely, $f(x) = ax^2 + bx + c$)"? Using the definition, one answer is to show that

$B1$: For all real numbers x, $f(x^*) \geq f(x)$.

Recognizing the quantifier "for all" in $B1$, the author uses the choose method to choose

$A1$: A real number x,

for which it must be shown that

$B2$: $f(x^*) \geq f(x)$, that is, that $a(x^*)^2 + bx^* + c \geq ax^2 + bx + c$.

The author does reach $B2$ by considering two separate cases.

Case 1. $x^* \geq x$. In this case, the author correctly notes that $x^* - x \geq 0$ and also that $a(x^* + x) + b \geq 0$ because $x^* = -b/(2a)$ and so

$A2$: $a(x^* + x) + b = -b/2 + ax + b = (2ax + b)/2$.

However, because $x^* = -b/(2a) \geq x$ and $a < 0$, it follows that

$A3$: $2ax + b \geq 0$.

Thus, from $A2$ and $A3$, the author correctly concludes that

$A4: a(x^* + x) + b \geq 0$.

Multiplying $A4$ through by $x^* - x \geq 0$ and rewriting yields

$A5: a(x^*)^2 - ax^2 + bx^* - bx \geq 0$.

Bringing all x-terms to the right and adding c to both sides yields $B2$, thus completing this case.

Case 2. $x^* < x$. In this case, the author leaves the following steps for you to create. Now $x^* - x < 0$ and also $a(x^* + x) + b \leq 0$ because

$A2: a(x^* + x) + b = -b/2 + ax + b = (2ax + b)/2$.

However, because $x^* = -b/(2a) < x$ and $a < 0$, it follows that

$A3: 2ax + b < 0$.

Thus, from $A2$ and $A3$,

$A4: a(x^* + x) + b < 0$.

Multiplying $A4$ through by $x^* - x < 0$ and rewriting yields

$A5: a(x^*)^2 - ax^2 + bx^* - bx > 0$.

Bringing all x-terms to the right and adding c to both sides yields $B2$, thus completing this case and the proof.

5.10 Analysis of Proof. Because the conclusion contains the key words "for all," the choose method is used to choose

$A1:$ Real numbers x and y with $x < y$,

for which it must be shown that

$B1: f(x) < f(y)$, that is,
$B2: mx + b < my + b$.

Work forward from the hypothesis that $m > 0$ to multiply both sides of the inequality $x < y$ in $A1$ by m yielding

$A2: mx < my$.

Adding b to both sides gives precisely $B2$.

5.13 Analysis of Proof. The forward-backward method gives rise to the key question, "How can I show that a function (namely, $x + 1$) is greater than or equal to another function (namely, $(x - 1)^2$) on a set (namely, S)?" The definition provides the answer that one must show that

$B1:$ For all $x \in S$, $x + 1 \geq (x - 1)^2$.

The appearance of the quantifier "for all" in the backward process suggests using the choose method to choose

A1 : An element $x \in S$,

for which it must be shown that

B2 : $x + 1 \geq (x - 1)^2$.

Bringing the term $x + 1$ to the right and performing algebraic manipulations, it must be shown that

B3 : $x(x - 3) \leq 0$.

However, working forward from A1 using the definition of S yields

A2 : $0 \leq x \leq 3$.

But then $x \geq 0$ and $x - 3 \leq 0$ and so B3 is true, thus completing the proof.

Proof. Let $x \in S$, so $0 \leq x \leq 3$. It will be shown that $x + 1 \geq (x - 1)^2$. However, because $x \geq 0$ and $x \leq 3$, it follows that $x(x-3) = x^2 - 3x \leq 0$. Adding $x + 1$ to both sides and factoring leads to the desired conclusion that $x + 1 \geq (x - 1)^2$. $\square$

5.16 Analysis of Proof. The forward-backward method gives rise to the key question, "How can I show a function is convex?" Using the definition in Exercise 5.1(e), one must show that

B1 : For all real numbers x and y, and for all t with $0 \leq t \leq 1$,
$f(tx + (1 - t)y) \leq tf(x) + (1 - t)f(y)$.

The appearance of the quantifier "for all," suggests using the choose method whereby one chooses

A1 : Real numbers x' and y', and a real number t' that satisfies $0 \leq t' \leq 1$,

for which it must be shown that

B2 : $f(t'x' + (1 - t')y') \leq t'f(x') + (1 - t')f(y')$.

But by the hypothesis,

$$
\begin{aligned}
A2 : \quad f(t'x' + (1 - t')y') &= m(t'x' + (1 - t')y') + b \\
&= mt'x' + my' - mt'y' + b \\
&= mt'x' + bt' + my' - mt'y' + b - bt' \\
&= t'(mx' + b) + (1 - t')(my' + b) \\
&= t'f(x') + (1 - t')f(y').
\end{aligned}
$$

Thus the desired inequality holds.

Proof. To show that f is convex, it will be shown that for all real numbers x and y, and for all t satisfying $0 \le t \le 1$, $f(tx + (1-t)y) \le tf(x) + (1-t)f(y)$. Let x' and y' be real numbers, and let t' satisfy $0 \le t' \le 1$, then

$$
\begin{aligned}
f(t'x' + (1-t')y') &= m(t'x' + (1-t')y') + b \\
&= mt'x' + my' - mt'y' + b \\
&= t'(mx' + b) + (1-t')(my' + b) \\
&= t'f(x') + (1-t')f(y').
\end{aligned}
$$

Thus the inequality holds and the proof is complete. $\square$

Solutions to Selected Exercises in Chapter 6

6.2 You can apply specialization to the statement,

> A1 : For every object X with the property Q, T happens,

to reach the conclusion that S happens for the particular object Y if, when you replace X with Y in A1, the something that happens, namely, T, becomes the desired conclusion that S happens for X. To apply specialization, you must show that the particular object Y has the property Q in A1, most likely by using the fact that Y has property P.

6.3 a. (1) Look for a specific real number, say, y, with which to apply specialization and (2) conclude that $f(y) \le f(x^*)$ as a new statement in the forward process.

b. (1) Look for a specific element, say, y, with which to apply specialization, (2) show that $y \in S$, and (3) conclude that $g(y) \ge f(y)$ as a new statement in the forward process.

c. (1) Look for a specific element, say, y, with which to apply specialization, (2) show that $y \in S$, and (3) conclude that $y \le u$ as a new statement in the forward process.

d. (1) Look for specific elements, say, x' and y', and a real number t' with which to apply specialization, (2) show that x', $y' \in C$ and that $0 \le t' \le 1$, and (3) conclude that $t'x' + (1-t')y' \in C$ as a new statement in the forward process.

e. (1) Look for specific real numbers, say, x', y', and t' with which to apply specialization, (2) show that $0 \le t' \le 1$, and (3) conclude that $f(t'x' + (1-t')y') \le t'f(x') + (1-t')f(y')$ as a new statement in the forward process.

6.5 a. To reach the desired conclusion that $\sin(2X) = 2\sin(X)\cos(X)$, specialize the given for-all statement to the angles $\alpha = X$ and $\beta = X$, which you can do because X is an angle. The result of this specialization is that $\sin(X + X) = \sin(X)\cos(X) + \cos(X)\sin(X)$, that is, $\sin(2X) = 2\sin(X)\cos(X)$.

b. To reach the desired conclusion that $(A \cap B)^c = A^c \cup B^c$, specialize the given for-all statement to the sets $S = A^c$ and $T = B^c$, which you can do because A^c and B^c are sets. The result of this specialization is that $(A^c \cup B^c)^c = (A^c)^c \cap (B^c)^c = A \cap B$. Applying the complement to both sides now leads to the desired conclusion that $A^c \cup B^c = (A \cap B)^c$.

6.7 **Analysis of Proof.** The forward-backward method gives rise to the key question, "How can I show that a set (namely, R) is a subset of another set (namely, T)?" One answer is by the definition, so one must show that

$B1$: For all $r \in R$, $r \in T$.

The appearance of the quantifier "for all" in the backward process suggests using the choose method. So choose

$A1$: An element $r' \in R$,

for which it must be shown that

$B2$: $r' \in T$.

Turning to the forward process, the hypothesis says that R is a subset of S and S is a subset of T. By definition, this means, respectively, that

$A2$: For all $r \in R$, $r \in S$, and
$A3$: For all $s \in S$, $s \in T$.

Specializing $A2$ to r' (which is in R from $A1$), one has that

$A4$: $r' \in S$.

Specializing $A3$ to r' (which is in S from $A4$), one has that

$A5$: $r' \in T$,

which is $B2$, thus completing the proof.

Proof. To show that $R \subseteq T$, it must be shown that for all $r \in R$, $r \in T$. Let $r' \in R$. By hypothesis, $R \subseteq S$, so $r' \in S$. Also, by hypothesis, $S \subseteq T$, so $r' \in T$. $\square$

6.9 **Analysis of Proof.** The forward-backward method gives rise to the key question, "How can I show that a set (namely, $S \cap T$) is convex?" One answer is by the definition, whereby it must be shown that

$B1$: For all $x, y \in S \cap T$, and for all t with $0 \le t \le 1$,
$tx + (1 - t)y \in S \cap T$.

The appearance of the quantifier "for all" in the backward process suggests using the choose method to choose

$A1$: $x', y' \in S \cap T$, and t' with $0 \le t' \le 1$, for which it must be shown that

B2 : $t'x' + (1 - t')y' \in S \cap T$.

Working forward from the hypothesis and $A1$, $B2$ will be established by showing that

B3 : $t'x' + (1 - t')y'$ is in both S and T.

Specifically, from the hypothesis that S is convex, by definition, it follows that

A2 : For all $x, y \in S$, and for all $0 \le t \le 1$, $tx + (1 - t)y \in S$.

Specializing $A2$ to $x = x'$, $y = y'$, and $t = t'$ (noting from $A1$ that $0 \le t' \le 1$) yields that

A3 : $t'x' + (1 - t')y' \in S$.

A similar argument shows that $t'x' + (1 - t')y' \in T$, thus completing the proof.

Proof. To see that $S \cap T$ is convex, let $x', y' \in S \cap T$, and let t' with $0 \le t' \le 1$. It will be established that $t'x' + (1 - t')y' \in S \cap T$. From the hypothesis that S is convex, one has that $t'x' + (1 - t')y' \in S$. Similarly, $t'x' + (1 - t')y' \in T$. Thus it follows that $t'x' + (1 - t')y' \in S \cap T$, and so $S \cap T$ is convex. □

6.13 The author has used the forward-backward method to ask the key question, "How can I show that a function (namely, g) is greater than or equal to another function (namely, f) on a set (namely, R)?" The definition provides the answer that it is necessary to show that

B1 : For every element $r \in R$, $g(r) \ge f(r)$.

Recognizing the quantifier "for every" in $B1$, the author uses the choose method to choose

A1 : An element $x \in R$,

for which it must be shown that

B2 : $g(x) \ge f(x)$.

To reach $B2$, the author turns to the forward process and works forward from the hypothesis that R is a subset of S by definition to claim that

A2 : For every element $r \in R$, $r \in S$.

Recognizing the quantifier "for every" in $A2$, the author specializes $A2$ to the element $r = x \in R$ chosen in $A1$. The result of this specialization is

$A3: x \in S$.

Then the author works forward from the hypothesis that $g \geq f$ on S, which, by definition, means that

$A4$: for every element $s \in S$, $g(s) \geq f(s)$.

Recognizing the quantifier "for every" in $A4$, the author specializes $A4$ to the element $s = x \in S$ (see $A3$). The result of this specialization is

$A5: g(x) \geq f(x)$.

The proof is now complete because $A5$ is the same as $B2$. □

6.15 The author makes a mistake when saying that, "In particular, for the specific elements x and y, and for the real number t, it follows that $tx + (1-t)y \in R$." In this sentence, the author is applying specialization to the statement

$A1$: For any two elements u and v in R, and for any real number
s with $0 \leq s \leq 1$, $su + (1-s)v \in R$.

Specifically, the author is specializing $A1$ with $u = x$, $v = y$, and $s = t$. However, the author has failed to verify that these specific objects satisfy the certain properties in $A1$. In particular, to specialize $A1$ to $u = x$ and $v = y$, it is necessary to verify that $x \in R$ and $y \in R$. The author fails to do so (in fact it is not necessarily true that $x \in R$ and $y \in R$). Thus, the author is not justified in applying this specialization.

Solutions to Selected Exercises in Chapter 7

7.1 a. For the quantifier "there is":
       Object: real number y.
       Certain property: none.
       Something happens: for every real number x, $f(x) \leq y$.
    For the quantifier "for every":
       Object: real number x.
       Certain property: none.
       Something happens: $f(x) \leq y$.

  b. For the quantifier "there is":
       Object: real number M.
       Certain property: $M > 0$.
       Something happens: $\forall$ elements $x \in S$, $|x| < M$.
    For the quantifier "for all":
       Object: element x.
       Certain property: $x \in S$.
       Something happens: $|x| < M$.

c. For the first quantifier "for all":
   Object: real number ϵ.
   Certain property: $\epsilon > 0$.
   Something happens: there is a real number $\delta > 0$ such that, for all real
                              numbers y with $|x - y| < \delta$, $|f(x) - f(y)| < \epsilon$.
   For the quantifier "there is":
   Object: real number δ.
   Certain property: $\delta > 0$.
   Something happens: for all real numbers y with $|x - y| < \delta$,
                              $|f(x) - f(y)| < \epsilon$.
   For the second quantifier "for all":
   Object: real number y.
   Certain property: $|x - y| < \delta$.
   Something happens: $|f(x) - f(y)| < \epsilon$.

d. For the first quantifier "for all":
   Object: real number ϵ.
   Certain property: $\epsilon > 0$.
   Something happens: $\exists$ an integer $k' \ni \forall$ integers k with $k > k'$,
                              $|x^k - x| < \epsilon$.
   For the quantifier "there is":
   Object: integer k'.
   Certain property: none.
   Something happens: $\forall$ integers k with $k > k'$, $|x^k - x| < \epsilon$.
   For the second quantifier "for all":
   Object: integer k.
   Certain property: $k > k'$.
   Something happens: $|x^k - x| < \epsilon$.

7.3 a. Both $S1$ and $S2$ are true. This is because when you apply the choose
         method to each statement, in either case you will choose real numbers
         x and y with $0 \le x \le 1$ and $0 \le y \le 2$ for which you can then show
         that $2x^2 + y^2 \le 6$.

b. $S1$ and $S2$ are different—$S1$ is true and $S2$ is false. You can use
   the choose method to show that $S1$ is true. To see that $S2$ is false,
   consider $y = 1$ and $x = 2$. For these real numbers, $2x^2 + y^2 = 2(4) + 1 = 9 > 6$.

c. These two statements are the same when the properties P and Q do
   not depend on the objects X and Y. This is the case in part (a) but
   not in part (b).

7.5 a. The construction method is used first to construct a real number
         $M > 0$. The choose method is used next to show that, for the value
         of M you constructed, it is true that for all elements $x \in T$, $|x| \le M$.
         In so doing, you would choose an element $x \in T$, for which you must
         then show that $|x| \le M$.

b. The choose method is used first to choose a real number $M > 0$, for which it must be shown that there is an element $x \in T$ such that $|x| > M$. To show this, the construction method is used next whereby you must construct an element $x \in T$ and then show that this element x satisfies $|x| > M$.

c. The choose method is used first to choose a real number $\epsilon > 0$ for which it must be shown that there is a real number $\delta > 0$ such that for all real numbers y with $|x - y| < \delta$, $|f(x) - f(y)| < \epsilon$. To show this, the construction method is used next to construct a real number $\delta > 0$. You must then show that this value of δ satisfies the property that for all real numbers y with $|x - y| < \delta$, $|f(x) - f(y)| < \epsilon$. To do this, use the choose method next to choose a real number y with $|x - y| < \delta$, for which you must show that $|f(x) - f(y)| < \epsilon$.

7.7 **Analysis of Proof.** The key words "for every" in the conclusion suggest using the choose method to choose

$A1$: A real number $x' > 2$,

for which it must be shown that

$B1$: There is a real number $y < 0$ such that $x' = 2y/(1 + y)$.

The key words "there is" in $B1$ suggest using the construction method to construct the desired y. Working backward form the fact that y must satisfy $x' = 2y/(1 + y)$, it follows that y must be constructed so that

$B2$: $x' + x'y = 2y$, or
$B3$: $y(2 - x') = x'$, or
$B4$: $y = x'/(2 - x')$.

To see that the value of y in $B4$ is correct, it is easily seen that $x' = 2y/(1 + y)$. However, it must also be shown that $y < 0$, which it is, because $x' > 2$.

Proof. Let $x' > 2$ and so it is possible to construct $y = x'/(2 - x')$. Because $x' > 2$, $y < 0$. It is also easy to verify that $x' = 2y/(1 + y)$. $\square$

Solutions to Selected Exercises in Chapter 8

8.1 a. Assume that l, m, and n are three consecutive integers, and that 24 divides $l^2 + m^2 + n^2 + 1$.

b. Assume that the matrix M is not singular and also that the rows of M are linearly dependent.

c. Assume that f and g are two functions such that $g \geq f$, f is unbounded above, and g is not unbounded above.

8.3 a. There are not a finite number of primes.

  b. The positive integer p cannot be divided by any positive integer other than 1 and p.

  c. The lines l and l' do not intersect.

8.5 a. Use the construction method to construct an element $s \in S$ and show that s is also in T.

  b. First use the choose method to choose an element $s \in S$, for which it must be shown that there is no element $t \in T$ such that $s > t$. To show the latter, use the contradiction method, whereby you should assume that s is an element of S and there is an element $t \in T$ such that $s > t$. Then you must work forward to reach a contradiction.

  c. First use the contradiction method, whereby you should assume that there is a real number $M > 0$ such that for all elements $x \in S$, $|x| < M$. To reach a contradiction, you will probably have to apply specialization to the statement, "for all elements $x \in S$, $|x| < M$."

8.7 **Analysis of Proof.** To use the contradiction method, assume:

$$A: \quad n \text{ is an integer for which } n^2 \text{ is even.}$$
$$NOT \ B: \quad n \text{ is not even, that is, } n \text{ is odd.}$$

Work forward from these assumptions using the definition of an odd integer to reach the contradiction that

$$B1: \ n^2 \text{ is odd.}$$

Applying a definition to work forward from $NOT\ B$ yields

$$A1: \text{ There exists an integer } k \text{ such that } n = 2k + 1.$$

Squaring both sides of the equality in $A1$ and performing simple algebraic manipulations leads to

$$A2: \ n^2 = (2k+1)^2, \text{ so}$$
$$A3: \ n^2 = 4k^2 + 4k + 1, \text{ so}$$
$$A4: \ n^2 = 2(2k^2 + 2k) + 1,$$

which says that $n^2 = 2p + 1$, where $p = 2k^2 + 2k$. Thus n^2 is odd, and this contradiction establishes the result.

Proof. Assume, to the contrary, that n is odd and n^2 is even. Hence, there is an integer k such that $n = 2k + 1$. Consequently,

$$
\begin{aligned}
n^2 &= (2k+1)^2 \\
&= 4k^2 + 4k + 1 \\
&= 2(2k^2 + 2k) + 1,
\end{aligned}
$$

and so n^2 is odd, contradicting the initial assumption. $\square$

8.9 **Analysis of Proof.** Proceed by assuming that there is a chord of a circle that is longer than its diameter. Using this assumption and the properties of a circle, you must arrive at a contradiction.

Let AC be the chord of the circle that is longer than the diameter of the circle (see the figure below). Let AB be a diameter of the circle. This construction is valid because, by definition, a diameter is a line passing through the center terminating at the perimeter of the circle.

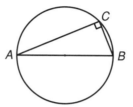

It follows that angle ACB has 90 degrees because ACB is an angle inscribed in a semicircle. Hence the triangle ABC is a right triangle in which AB is the hypotenuse. Then the desired contradiction is that the hypotenuse of a right triangle is shorter than one side of the triangle, which is impossible.

Proof. Assume that there does exist a chord, say, AC, of a circle that is longer than a diameter. Construct a diameter that has one of its ends coinciding with one end of the chord AC. Joining the other ends produces a right triangle in which the diameter AB is the hypotenuse. But then the hypotenuse is shorter than one leg of the right triangle, which is a contradiction. □

8.13 **Analysis of Proof.** To use contradiction, assume that no two people have the same number of friends, that is, everybody has a different number of friends. Because there are n people, each of whom has a different number of friends, you can number the people in an increasing sequence according to the number of friends that person has. In other words,

person number 1 has no friends,
person number 2 has 1 friend,
person number 3 has 2 friends,

$\vdots$

person number n has $n - 1$ friends.

By doing so, there is a contradiction that the last person is friends with all the other $n - 1$ people, including the first one, who has no friends.

Proof. Assume, to the contrary, that no two people have the same number of friends. Number the people at the party in such a way that

person number 1 has no friends,
person number 2 has 1 friend,
person number 3 has 2 friends,

$$\vdots$$

person number n has $n - 1$ friends.

It then follows that the person with $n-1$ friends is a friend of the person who has no friends, which is a contradiction. ☐

8.16 The contradiction is that b is both odd (as stated in the hypothesis) and even (as has been shown because $b = -2a$ is even).

8.18 Analysis of Proof. The proof is done by contradiction because the author is assuming

$$NOT\ B:\ xz - y^2 > 0.$$

A contradiction is reached by showing that the square of a number is negative. Specifically, the author shows that

$$B1:\ (az - cx)^2 < 0.$$

The contradiction in $B1$ is obtained by working forward from $NOT\ B$ and the hypothesis. Specifically, from $NOT\ B$ and the hypothesis that $ac - b^2 > 0$, it follows that

$$A1:\ xz > y^2,$$
$$A2:\ ac > b^2.$$

Multiplying corresponding sides of the inequalities in $A1$ and $A2$ yields:

$$A3:\ (ac)(xy) > b^2y^2.$$

Then, adding $2by$ to both sides of the hypothesis $az - 2by + cx = 0$ and squaring both sides results in:

$$A4:\ (az + cx)^2 = 4b^2y^2.$$

From $A3$, it follows that $4b^2y^2 < 4(ac)(xy)$ and so, from $A4$,

$$A5:\ (az + cx)^2 < 4(ac)(xy).$$

Expanding the left side of $A5$, subtracting $4(ac)(xy)$ from both sides, and then factoring yields the contradiction in $B1$ that $(az - cx)^2 < 0$.
☐

Solutions to Selected Exercises in Chapter 9

9.1 a. Work forward from: n is an odd integer.
Work backward from: n^2 is an odd integer.

b. Work forward from: S is a subset of T and T is bounded.
Work backward from: S is bounded.

c. Work forward from: The rows of M are linearly dependent.
Work backward from: M is singular.

9.3 Statement (b) is a result of the forward process because you can assume that there is a real number t with $0 < t < \pi/4$ such that $\sin(t) = r\cos(t)$. The answer in part (b) results on squaring both sides and replacing $\cos^2(t)$ with $1 - \sin^2(t)$.

9.5 With the contrapositive method you work forward from $NOT\ B$ and backward from $NOT\ A$. Thus, you should apply the key question to the statement, "The derivative of the function f at the point x is 0."

a. Incorrect because the key question is applied to $NOT\ B$. Also, the question uses symbols and notation from the specific problem.

b. Incorrect because the key question uses symbols and notation from the specific problem.

c. Incorrect because the key question is applied to $NOT\ B$.

d. Correct.

9.7 **Analysis of Proof.** By the contrapositive method, assume that

$$NOT\ B:\ p = q.$$

It must be shown that

$$NOT\ A:\ \sqrt{pq} = (p + q)/2.$$

However, from $NOT\ B$ and the fact that $p,\ q > 0$, it follows that

$$A1:\ \sqrt{pq} = \sqrt{p^2} = p = (p + p)/2 = (p + q)/2.$$

The proof is now complete because $NOT\ A$ is true.

Proof. By the contrapositive method, assume that $p = q$. It must be shown that $\sqrt{pq} = (p + q)/2$. However, because $p,\ q > 0$, it follows that

$$\sqrt{pq} = \sqrt{p^2} = p = (p + p)/2 = (p + q)/2.$$

The proof is now complete. □

9.13 Analysis of Proof. This is a proof by the contrapositive method because the author assumes that

$$NOT\ B:\ p \text{ is not prime}$$

and eventually shows that

NOT A: There is an integer m with $1 < m \le \sqrt{p}$ such that $m|p$.

Recognizing the key words "there is" in NOT A, the construction method is used to produce the value for the integer m. To do so, the author works forward from NOT B to claim that, because p is not prime,

$A1$: There is an integer n with $1 < n < p$ such that $n|p$.

The author now considers two possibilities: $n \le \sqrt{p}$ and $n > \sqrt{p}$. In the former case,

$A2$: $n \le \sqrt{p}$,

and the author constructs

$A3$: $m = n$.

This value for m in $A3$ is correct because, from $A1$ and $A2$, $1 < n = m$ and $m = n \le \sqrt{p}$. Also, from $A1$, $n|p$ and $m = n$, so $m|p$.

In the latter case,

$A4$: $n > \sqrt{p}$.

To construct the value for m, the author works forward from $A1$ and the definition of $n|p$ to state that

$A5$: There is an integer k such that $p = nk$.

The author then constructs

$A6$: $m = k$.

The author shows that this value for m is correct by establishing that

$A7$: $1 < k \le \sqrt{p}$.

To do so, the author argues by contradiction that $1 < k$ for otherwise, from $A5$, it would follow that $p = nk \le n$, which cannot happen because, from $A1$, $n < p$. Finally, $k \le \sqrt{p}$ because otherwise, $k > \sqrt{p}$ and then, since $n > \sqrt{p}$ from $A4$, it would follow that $nk > \sqrt{p}\sqrt{p} = p$, which cannot happen because, from $A5$, $p = nk$. Observe that the author omits noting that $m|p$, which is true because, from $A5$, $p = nk = mk$. The proof is now complete. $\square$

Solutions to Selected Exercises in Chapter 10

10.1 a. The real number x^* is not a maximum of the function f means that there is a real number x such that $f(x) > f(x^*)$.

b. Suppose that f and g are functions of one variable. Then g is not $\geq f$ on the set S of real numbers means that there exists an element $x \in S$ such that $g(x) < f(x)$.

c. The real number u is not an upper bound for a set S of real numbers means that there exists an $x \in S$ such that $x > u$.

d. The set C of real numbers is not a convex set means that there exist elements $x, y \in C$ and there exists a real number t with $0 \leq t \leq 1$ such that $tx + (1 - t)y \notin C$.

e. The function f of one variable is not convex means that there exist real numbers x and y and t with $0 \leq t \leq 1$ such that $f(tx + (1-t)y) > tf(x) + (1 - t)f(y)$.

10.3 a. There does not exist an element $x \in S$ such that $x \notin T$.

b. It is not true that for every angle t between 0 and $\pi/2$, $\sin(t) \neq \cos(t)$.

c. There does not exist an object with the certain property such that the something does not happen.

d. It is not true that for every object with the certain property, the something does not happen.

10.5 a. Work forward from $NOT\ B$.
Work backward from $(NOT\ A)\ OR\ (NOT\ C)$.

b. Work forward from $NOT\ B$.
Work backward from $(NOT\ A)\ AND\ (NOT\ C)$.

c. Work forward from $(mn$ is not divisible by 4) AND (n is divisible by 4).
Work backward from (n is an odd integer) OR (m is an even integer).

Solutions to Selected Exercises in Chapter 11

11.1 Analysis of Proof. The direct uniqueness method is used, whereby one must first construct the real number y. This is done in Exercise 7.7. It remains to show the uniqueness by assuming that y and z are real numbers with

$$A1: \quad y < 0.$$
$$A2: \quad z < 0.$$
$$A3: \quad x = 2y/(1 + y).$$
$$A4: \quad x = 2z/(1 + z).$$

Working forward via algebraic manipulations, it will be shown that

$$B1: \quad y = z.$$

Specifically, combining $A3$ and $A4$ yields

$$A5: \quad x = 2y/(1 + y) = 2z/(1 + z).$$

Dividing both sides of $A5$ by 2 and clearing the denominators, you have that

$A6: y + yz = z + yz.$

On subtracting yz from both sides, the desired conclusion that $y = z$ follows.

Proof. The existence of the real number y is established in Exercise 7.7. To show that y is unique, suppose that y and z satisfy $y < 0$, $z < 0$, $x = 2y/(1+y)$, and also $x = 2z/(1+z)$. But then $2y/(1+y) = 2z/(1+z)$, and so $y + yz = z + yz$, or, $y = z$, as desired. $\square$

11.5 Analysis of Proof. The issue of existence is addressed first. To construct the desired complex number $c+di$ that satisfies $(a+bi)(c+di) = 1$, multiply the two terms using complex arithmetic to obtain

$$ac - bd = 1, \quad \text{and}$$
$$bc + ad = 0.$$

Solving these two equations for the two unknowns c and d in terms of a and b leads you to construct $c = a/(a^2 + b^2)$ and $d = -b/(a^2 + b^2)$ (noting that the denominator is not 0 because, by the hypothesis, at least one of a and b is not 0). To see that this construction is correct, note that

$$
\begin{aligned}
A1: (a + bi)(c + di) &= (ac - bd) + (bc + ad)i \\
&= [(a^2 + b^2)/(a^2 + b^2)] + 0i \\
&= 1.
\end{aligned}
$$

To establish the uniqueness, suppose that $e + fi$ is another complex number that also satisfies

$A2: (e + fi)(a + bi) = 1.$

It will be shown that

$B1: c + di = e + fi.$

Working forward by multiplying both sides of the equality in $A1$ by $e + fi$ and using associativity yields

$A3: [(e + fi)(a + bi)](c + di) = (e + fi).$

Because $(e + fi)(a + bi) = 1$ from $A2$, it follows from $A3$ that $c + di = e + fi$, and so $B1$ is true, completing the proof.

Proof. Because either $a \neq 0$ or $b \neq 0$, $a^2 + b^2 \neq 0$, and so it is possible to construct the complex number $c + di$ in which $c = a/(a^2 + b^2)$ and $d = -b/(a^2 + b^2)$, for then

$$(a + bi)(c + di) = ac - bd + (bc + ad)i = 1.$$

To see the uniqueness, suppose that $e + fi$ also satisfies $(a+bi)(e+fi) = 1$. Multiplying the foregoing displayed equality through by $e + fi$ yields

$$[(e + fi)(a + bi)](c + di) = e + fi.$$

Using the fact that $(a + bi)(e + fi) = 1$, it follows that $c + di = e + fi$ and so the uniqueness is established. $\square$

11.6 a. Applicable.

b. Not applicable because the statement contains the quantifier "there is" instead of "for all."

c. Applicable.

d. Applicable.

e. Not applicable because in this statement, n is a real number, and induction is applicable only to integers.

11.10 Proof. First it must be shown that $P(n)$ is true for $n = 1$. Replacing n by 1, it must be shown that $1(1!) = (1 + 1)! - 1$. But this is clear because $1(1!) = 1 = (1 + 1)! - 1$.

Now assume that $P(n)$ is true and use that fact to show that $P(n+1)$ is true. So assume

$$P(n) : \ 1(1!) + \cdots + n(n!) = (n + 1)! - 1.$$

It must be shown that

$$P(n + 1) : \ 1(1!) + \cdots + (n + 1)(n + 1)! = (n + 2)! - 1.$$

Starting with the left side of $P(n + 1)$ and using $P(n)$:

$$
\begin{aligned}
1(1!) + &\cdots + n(n!) + (n + 1)(n + 1)! \\
&= \ [1(1!) + \cdots + n(n!)] + (n + 1)(n + 1)! \\
&= \ [(n + 1)! - 1] + (n + 1)(n + 1)! \\
&= \ (n + 1)![1 + (n + 1)] - 1 \\
&= \ (n + 1)!(n + 2) - 1 \\
&= \ (n + 2)! - 1. \ \square
\end{aligned}
$$

11.14 Proof. The statement is true for a set consisting of one element, say, x, because its subsets are $\{x\}$ and $\emptyset$, that is, there are $2^1 = 2$ subsets. Assume that for a set with n elements, the number of subsets is 2^n. It will be shown that for a set with $n+1$ elements, the number of subsets is 2^{n+1}. For a set S with $n + 1$ elements, one can construct all the subsets by listing first all those subsets that include the first n elements, and then, to each such subset, one can add the last element of S. By the induction hypothesis, there are 2^n subsets using the first n elements. An additional 2^n subsets are created by adding the last element of S to each of the subsets of n elements. Thus the total number of subsets of S is $2^n + 2^n = 2^{n+1}$, and so the statement is true for $n + 1$. $\square$

11.16 Proof. Let

$$S = 1 + 2 + \cdots + n.$$

Then

$$S = n + (n-1) + \cdots + 1.$$

On adding the two foregoing equations, one obtains

$$2S = n(n+1), \text{ that is, } S = n(n+1)/2. \quad \square$$

11.20 a. Verify that $P(n)$ is true for the initial value of $n = n_0$. Then, assuming that $P(n)$ true, prove that $P(n-1)$ is true.

b. Verify that $P(n)$ is true for some integer n_0. Assuming that $P(n)$ is true for n, prove that $P(n+1)$ is true and that $P(n-1)$ is also true.

11.22 The author relates $P(n+1)$ to $P(n)$ by expressing the product of the n terms associated with $P(n+1)$ as the product of the $n-1$ terms associated with $P(n)$ and one additional term—specifically,

$$\underbrace{\prod_{k=2}^{n+1}\left(1 - \frac{1}{k^2}\right)}_{P(n+1)} = \underbrace{\left[\prod_{k=2}^{n}\left(1 - \frac{1}{k^2}\right)\right]}_{P(n)}\underbrace{\left(1 - \frac{1}{(n+1)^2}\right)}_{\text{extra term}}$$

The induction hypothesis is used when the author subsequently replaces $\prod_{k=2}^{n}(1 - \frac{1}{k^2})$, which is the left side of $P(n)$, with $\frac{n+1}{2n}$, which is the right side of $P(n)$.

11.25 The mistake occurs in the last sentence where it states that, "Then, because all the colored horses in this (second) group are brown, the uncolored horse must also be brown." How do you know that there is a colored horse in the second group? In fact, when the original group of $n+1$ horses consists of exactly 2 horses, the second group of n horses does not contain a colored horse. The entire difficulty is caused by the fact that the statement should have been verified for the initial integer $n = 2$, not $n = 1$. This, of course, you will not be able to do.

Solutions to Selected Exercises in Chapter 12

12.1 a. To apply a proof by elimination to the statement, "If A, then C OR D OR E," you would assume that A is true, C is not true, and D is not true; you must conclude that E is true. (Alternatively, you can assume that A is true and that any two of the three statements C, D, and E are not true; you would then have to conclude that the remaining statement is true.)

b. To apply a proof by cases to the statement, "If C OR D OR E, then B," you must do all three of the following proofs: (1) "If C, then B," (2) "If D, then B," and (3) "If E, then B."

12.3 A proof by cases is used in the proof of Proposition 20 when the author tries to show that it is possible to express the integer $n+1$ as the product of primes. In particular, the author considers the following two cases because, in the first case, the desired conclusion follows immediately:

Case 1: $n+1$ is prime. In this case, $n+1$ is the product of primes, namely, itself.

Case 2: $n+1$ is not prime. In this case, the author uses this fact to express $n+1$ as the product of a prime p and an integer q to which the induction hypothesis can be applied to complete the proof.

12.7 a. If x is a real number that satisfies $x^3 + 3x^2 - 9x - 27 \geq 0$, then $x \leq -3$ or $x \geq 3$.

b. **Analysis of Proof.** According to the either/or method, you can assume that

$$A: \quad x^3 + 3x^2 - 9x - 27 \geq 0, \text{ and}$$
$$A1\ (NOT\ C): \quad x > -3.$$

It must be shown that

$$B1\ (D): \quad x \geq 3, \text{ that is, } x - 3 \geq 0.$$

By factoring A, it follows that

$$A2: \quad x^3 + 3x^2 - 9x - 27 = (x-3)(x+3)^2 \geq 0.$$

From $A1$, because $x > -3$, $(x+3)^2$ is strictly positive. Thus, dividing both sides of $A2$ by $(x+3)^2$ yields $B1$ and completes the proof.

Proof. Assume that $x^3 + 3x^2 - 9x - 27 \geq 0$ and $x > -3$. Then it follows that

$$x^3 + 3x^2 - 9x - 27 = (x-3)(x+3) \geq 0.$$

Because $x > -3$, $(x+3)^2$ is positive, so $x - 3 \geq 0$, or equivalently, $x \geq 3$. $\square$

12.11 Analysis of Proof. The appearance of the key words either/or in the hypothesis suggest proceeding with a proof by cases.

Case 1. Assume that

$$A1: \quad a | b.$$

It must be shown that

B1 : $a|(bc)$.

B1 gives rise to the key question, "How can I show that an integer (namely, a) divides another integer (namely, bc)?" Applying the definition means you must show that

B2 : There is an integer k such that $bc = ka$.

Recognizing the key words "there is" in B2, you should use the construction method to produce the desired integer k. Working forward from A1 by definition, you know that

A2 : There is an integer p such that $b = pa$.

Multiplying both sides of the equality in A2 by c yields

A3 : $bc = cpa$.

From A3, it is easy to see that the desired value for k is cp, thus completing this case.

Case 2. In this case, you should assume that

A1 : $a|c$.

You must show that

B1 : $a|(bc)$.

The remainder of the proof in this case is similar to that in Case 1 and is not repeated.

Proof. Assume, without loss of generality, that $a|b$. By definition, there is an integer p such that $b = pa$. But then, $bc = (cp)a$, and so $a|(bc)$. $\square$

12.13 a. For all $s \in S$, $s \leq x$.

   b. There is an $s \in S$ such that $s \geq x$.

   c. There is an x with $ax \leq b$ and $x \geq 0$ such that $cx \leq u$.

   d. There is an x with $ax \leq b$ and $x \geq 0$ such that $cx \geq u$.

   e. For all x with $b \leq x \leq c$, $ax \geq u$.

   f. For all x with $b \leq x \leq c$, $ax \leq u$.

12.15 Analysis of Proof. Recognizing the key word "min" in the conclusion of the proposition and letting $z = \max\{ub : ua \leq c, u \geq 0\}$, the max/min methods results in the following equivalent quantified statement:

$B1$: For every real number x with $ax \geq b$ and $x \geq 0$, it follows that $cx \geq z = \max\{ub : ua \leq c, u \geq 0\}$.

The key words "for every" in $B1$ suggest using the choose method to choose

$A1$: A real number x with $ax \geq b$ and $x \geq 0$,

for which it must be shown that

$B2$: $cx \geq \max\{ub : ua \leq c, u \geq 0\}$.

Recognizing the key word "max" in $B2$, the max/min methods lead to the need to prove the following equivalent quantified statement:

$B3$: For every real number u with $ua \leq c$ and $u \geq 0$, $cx \geq ub$.

The key words "for every" in $B3$ suggest using the choose method to choose

$A2$: A real number u with $ua \leq c$ and $u \geq 0$,

for which it must be shown that

$B4$: $cx \geq ub$.

To reach $B4$, multiply both sides of $ax \geq b$ in $A1$ by $u \geq 0$ to obtain

$A3$: $uax \geq ub$.

Likewise, multiply both sides of $ua \leq c$ in $A2$ by $x \geq 0$ to obtain

$A4$: $uax \leq cx$.

The desired conclusion in $B4$ that $cx \geq ub$ follows by combining $A3$ and $A4$.

Proof. To reach the desired conclusion that $cx \geq ub$, let x and u be real numbers with $ax \geq b$, $x \geq 0$, $ua \leq c$, and $u \geq 0$. Multiplying $ax \geq b$ through by $u \geq 0$ and $ua \leq c$ through by $x \geq 0$, it follows that

$$ub \leq uax \leq cx. \quad \Box$$

Solutions to Selected Exercises in Chapter 13

13.1 a. Contrapositive or contradiction method because the key word "no" appears in the conclusion.

b. Induction method because the conclusion is true for every integer $n \geq 4$.

c. Forward-Backward method because there are no key words in the hypothesis or conclusion.

d. Max/Min method because the conclusion contains the key word "maximum."

e. Uniqueness method because the conclusion contains the key words "one and only one."

f. Contradiction or contrapositive method because "no" is the first key word to appear in the conclusion.

g. Forward-Backward method because there are no key words in the hypothesis and conclusion.

h. Choose method because the first quantifier from the left in the conclusion is "for every."

i. Construction method because the first quantifier from the left in the conclusion is "there is."

Glossary of Math Terms and Symbols

and—For two statements A and B, the statement A *AND* B (written $A \wedge B$) is true when both A and B are true, and false otherwise.

axiom—A statement whose truth is accepted without a proof.

backward process—The process of deriving, from a statement B, a new statement, $B1$, with the property that if $B1$ is true, then so is B. You do this by asking and answering a key question.

bounded above function—A function f of one variable for which there is a real number y such that for every real number x, $f(x) \leq y$.

bounded set of real numbers—A set of real numbers S for which there is a real number $M > 0$ such that for every element $x \in S$, $|x| < M$.

choose method—A technique for proving that for every object with a certain property, something happens. To do so, choose a generic object with the certain property. Then show that, for this chosen object, the something happens.

conclusion—The statement B in the implication "A implies B." When proving "A implies B," your job is to show that the conclusion B is true.

conditional statement—A statement whose truth depends on the value of one or more variables.

construction method—A technique for proving that there is an object with a certain property such that something happens. To do so, construct, guess, produce, or devise an algorithm to produce the desired object. Then show that the object you constructed has the certain property and satisfies the something that happens.

continuous function at a point—A function f of one variable such that, at a given point x, for every real number $\epsilon > 0$, there is a real number $\delta > 0$ such that, for all real numbers y with $|x - y| < \delta$, $|f(x) - f(y)| < \epsilon$.

contradiction method—A technique for proving that "A implies B" in which you work forward from the assumption that A and $NOT\ B$ are true to reach a contradiction to some statement that you know is true.

contrapositive method—A technique for proving that "A implies B" in which you prove that "$NOT\ B$ implies $NOT\ A$," that is, you work forward from $NOT\ B$ and backward from $NOT\ A$.

contrapositive statement—The contrapositive of the statement "A implies B" is the statement "$NOT\ B$ implies $NOT\ A$."

convergence of a sequence of real numbers to a given real number— A sequence of real numbers $x_1, x_2, \ldots$ for which, at a given real number x, for every real number $\epsilon > 0$, there is an integer $k' \geq 1$ such that for every integer k with $k > k'$, $|x^k - x| < \epsilon$.

converse statement—The converse of the statement "A implies B" is the statement "B implies A."

convex function— A function f of one variable such that, for all real numbers x and y and for all real numbers t with $0 \leq t \leq 1$, it follows that $f(tx + (1 - t)y) \leq tf(x) + (1 - t)f(y)$.

convex set—A set C of real numbers such that, for all elements $x, y \in C$ and for all real numbers t with $0 \leq t \leq 1$, $tx + (1 - t)y \in C$.

corollary—A proposition whose truth follows almost immediately from a theorem.

decreasing sequence of real numbers—A sequence $x_1, x_2, \ldots$ of real numbers such that for every integer $k = 1, 2, \ldots$, $x_k > x_{k+1}$.

definition—An agreement, by all parties concerned, as to the meaning of a particular term.

derivative of a function at a point—The real number $f'(\bar{x})$ is the derivative of the function f at the point $\bar{x}$ if and only if for every real number $\epsilon > 0$, there is a real number $\delta > 0$ such that for every real number x with $0 < |x - \bar{x}| < \delta$, it follows that $|f(x) - f(\bar{x})|/|x - \bar{x}| < \epsilon$.

direct uniqueness method—A technique for proving that there is a unique object with a certain property such that something happens. To do so, first construct the desired object X. Then assume that there is also an object Y with the certain property and for which the something happens. Work forward to show that X and Y are the same.

divides—An integer a divides an integer b (written $a|b$) if and only if there is an integer c such that $b = ca$.

either/or methods—Techniques for proving that "A implies B" when A and/or B contain the key words *either/or* (see proof by elimination and proof by cases).

element of a set—An item that belongs to a given set.

empty set—The set with no elements, written $\emptyset$.

equal pairs of real numbers—Two pairs of real numbers (x_1, y_1) and (x_2, y_2) for which $x_1 = x_2$ and $y_1 = y_2$.

equal sets—Two sets S and T are equal (written $S = T$) if and only if S is a subset of T and T is a subset of S.

equilateral triangle—A triangle all of whose sides have the same length.

equivalent statements—Two statements A and B for which "A implies B" and "B implies A."

even integer—An integer n whose remainder on dividing by 2 is 0. Equivalently, an integer n for which there is an integer k such that $n = 2k$.

existential quantifier—The key words *there is (there are, there exists)*.

forward-backward method—The technique for proving that "A implies B" in which you assume that A is true and try to show that B is true. To do so, you apply the forward process to A and the backward process to B.

forward process—The process of deriving from a statement A, a new statement, $A1$, with the property that $A1$ is true because A is true.

greater than or equal to functions—For functions f and g of one variable, $g \geq f$ on the set S of real numbers if and only if for every element $x \in S$, $g(x) \geq f(x)$.

greatest common divisor—An integer d such that, for two given integers a and b, (1) d divides a and d divides b and (2) whenever c is an integer for which c divides a and c divides b, it follows that c divides d.

hypothesis—The statement A in the implication "A implies B." When proving "A implies B," you can assume that the hypothesis A is true.

implication—A statement of the form, "If A is true, then B is true," where A and B are given statements.

increasing sequence of real numbers—A sequence $x_1, x_2, \ldots$ of real numbers such that for every integer $k = 1, 2, \ldots$, $x_k < x_{k+1}$.

indirect uniqueness method—A technique for proving that there is a unique object with a certain property such that something happens. To do so, first construct the desired object X. Then assume that there is an object Y, different from X, with the certain property and for which the something happens. Work forward to reach a contradiction.

induction—A technique for proving that for every integer $n \geq$ some initial integer n_0, some statement $P(n)$ is true. To do so, show that $P(n_0)$ is true.

Then assume that $P(n)$ is true and show that $P(n+1)$ is true by relating $P(n+1)$ to $P(n)$.

inverse statement—The inverse of the statement "A implies B" is the statement "NOT A implies NOT B."

isosceles triangle—A triangle, two of whose sides have equal length.

key answer—An answer to a key question.

key question—The specific question obtained by asking how you can show that a given statement B is true.

least upper bound for a set—A real number u such that, for a given set S of real numbers, (1) u is an upper bound for S and (2) for every upper bound v for S, $u \leq v$.

lemma—A proposition that is used in the proof of a subsequent theorem.

maximum of a function—A real number x^* such that for every real number x, $f(x) \leq f(x^*)$ (where f is a function of one variable).

max/min methods— Methods for proving that the maximum or minimum of a given set S of real numbers is $\leq$ or $\geq$ a given real number x by converting the statement to an equivalent statement containing the quantifier *there is* or *for all*, as follows:

- $\min\{s \in S\} \geq x$ is equivalent to for all $x \in S$, $x \geq s$.

- $\min\{s \in S\} \leq x$ is equivalent to there is an $x \in S$ such that $x \leq s$.

- $\max\{s \in S\} \geq x$ is equivalent to there is an $x \in S$ such that $x \geq s$.

- $\max\{s \in S\} \leq x$ is equivalent to for all $x \in S$, $x \leq s$.

member of a set—An item that belongs to a given set.

nested quantifiers—A statement that contains more than one quantifier.

not—For a statement A, the statement NOT A is true when A is false and is false when A is true.

odd integer—An integer n for which there is an integer k such that $n = 2k + 1$.

one-to-one function—A function f of one variable such that for all real numbers x and y with $x \neq y$, $f(x) \neq f(y)$.

onto function—A function f of one variable such that for every real number y, there is a real number x such that $f(x) = y$.

or—For two statements A and B, the statement A OR B (written $A \bigvee B$) is false when A is false and B is false, and true otherwise.

proposition—A true statement of interest.

prime—An integer $p > 1$ whose only positive integer divisors are 1 and p.

proof—A convincing argument expressed in the language of mathematics.

proof by cases—A technique for proving that "A implies B" when A has the form "either C or D." To do so, first assume that C is true and show that B is true. Then assume that D is true and show that B is true.

proof by elimination—A technique for proving that "A implies B" when B has the form "either C or D." To do so, assume that A and NOT C are true and show that D is true. Alternatively, you can assume that A and NOT D are true and show that C is true.

proof technique—Any method for proving that the statement "A implies B" is true.

quantifier—One of the two groups of key words *there is (there are, there exists)* and *for all (for each, for every, for any)*.

rational number—A real number r for which there are integers p and q with $q \neq 0$ such that $r = p/q$.

set—A collection of items.

square integer—An integer n for which there is an integer k such that $n = k^2$.

standard form for "for all"—For all objects with a certain property, something happens.

standard form for "there is"—There is an object with a certain property such that something happens.

statement—A mathematical sentence that is either true or false.

strictly increasing function—A function f of one variable such that for all real numbers x and y with $x < y$, $f(x) < f(y)$.

strictly monotone sequence of real numbers—A sequence of real numbers $x_1, x_2, \ldots$ that is either decreasing or increasing.

subset—A set S is a subset of a set T (written $S \subseteq T$) if and only if for every element $x \in S$, $x \in T$.

theorem—An important proposition.

truth table—A table that lists the truth of a complex statement (such as "A implies B") for all possible combinations of truth values of the individual statements (in this case, A and B).

uniqueness methods—Techniques for proving that there is a unique object with a certain property such that something happens (see the direct and indirect uniqueness methods).

universal quantifier—The key words *for all (for each, for every, for any)*.

upper bound for a set—A real number u such that for all $x \in S$, $x \leq u$ (where S is a given set of real numbers).

Glossary of Mathematical Symbols

Symbol	Meaning	Page
$\Rightarrow$	implies	3
$\Leftrightarrow$	if and only if	25
$\in$	is an element of	45
$\subseteq$	subset	46
$\emptyset$	empty set	46
$\sim$	not	29
$\forall$	for all (for each, for any, for every)	47
$\exists$	there is (there are, there exists)	36
$\ni$	such that	36
$\wedge$	and	24
$\vee$	or	24
$\square$	Q. E. D. (which was to be demonstrated)	14

References

1. Avelsgaard, C. *Foundations for Advanced Mathematics*. Glenview, IL: Scott, Foresman, 1990.

2. Bittinger, Marvin L. *Logic, Proof, and Sets* (2 ed.). Reading, MA: Addison-Wesley, 1970.

3. Cupillari, Antonella. *The Nuts and Bolts of Proofs*. Belmont, CA: Wadsworth, 1989.

4. D'Angelo, J. P. and D. B. West. *Mathematical Thinking: Problem-Solving and Proofs* (2 ed.). Englewood Cliffs, NJ: Prentice-Hall, 2000.

5. Exner, George R. *An Accompaniment to Higher Mathematics*. New York, NY: Springer-Verlag, 1996.

6. Fendel, D. and D. Resek. *Foundations of Higher Mathematics*. Reading, MA: Addison-Wesley, 1970.

7. Fletcher, Peter and C. Wayne Patty. *Foundations of Higher Mathematics*. Boston, MA: PWS-Kent, 1988.

8. Gerstein, L. *Introduction to Mathematical Structures and Proofs*. New York, NY: Springer-Verlag, 1996.

9. Granier, Rowan, and John Taylor. *100% Mathematical Proof*. New York, NY: John Wiley, 1996.

10. Levine, Alan. *Discovering Higher Mathematics—Four Habits of Highly Effective Mathematicians.* San Diego, CA: Harcourt Academic Press, 2000.

11. Lucas, J. F. *Introduction to Abstract Mathematics.* Belmont, CA: Wadsworth, 1986.

12. Morash, R. P. *Bridge to Abstract Mathematics.* New York, NY: Random House, 1987.

13. Polya, George. *How to Solve It.* Garden City, NJ: Doubleday, 1957.

14. Polya, George. *Mathematical Discovery* (combined ed.). New York, NY: John Wiley, 1981.

15. Schumacher, Carol. *Chapter 0—Fundamental Notions of Abstract Mathematics.* Reading, MA: Addison-Wesley, 1996.

16. Smith, Douglas, Eggen, Maurice, and Richard St. Andree. *A Transition to Advanced Mathematics*, Monterey, CA: Brooks/Cole, 1983.

17. Solow, Daniel. *The Keys to Advanced Mathematics.* Cleveland, OH: Books Unlimited, 1995.

18. Solow, Daniel. *The Keys to Linear Algebra.* Cleveland, OH: Books Unlimited, 1998.

19. Wicklegren, Wayne A. *How to Solve Problems.* San Francisco, CA: W. H. Freeman, 1974.

Index

A implies *B*, 3, 30
alternative definition, 25
AND, 24, 100
answering a key question, 10
axiom, 29
B implies *A*, 30
backward process, 10
certain property, 36
choose method, 48, 53
conclusion, 3
condensed proof, 3, 13
    reading, 15
conditional statement, 2
construction method, 37, 40
contradiction method, 80, 86
contrapositive
    method, 91, 96
    statement, 30
    truth table, 30
converse statement, 30
corollary, 29
defining property of a set, 46
definition, 23
    alternative, 25
direct uniqueness method, 105, 114
doing proofs, 3
either/or methods, 121, 128
element of a set, 45
elimination method, 121, 128
empty set, 46

equal
    two sets, 47
equivalent
    definitions, 25
    statements, 24–25
existential quantifier, 35
    standard form, 36
for all ("for any" "for each" "for every"),
        35, 45
    standard form, 47
forward process, 10, 12, 16
forward-backward method, 9, 16
generalized induction, 112
hidden quantifier, 37, 47
hypothesis, 3
    induction, 110
if and only if, 25
iff, 25
implication, 3
implies, 3
indirect uniqueness method, 107, 114
induction, 108, 115
    generalized, 112
    using the induction hypothesis, 110
infinite
    list, 48
    set, 46
inverse statement, 30
key question, 10
lemma, 29

matching up notation, 27–28, 62
mathematical
  proof, 3
  terminology, 29
max/min methods, 126, 129
member of a set, 45
methods, 133
  cases, 122, 129
  choose, 48, 53
  construction, 37, 40
  contradiction, 80, 86
  contrapositive, 91, 96
  direct uniqueness, 105
  either/or, 121, 128
  elimination, 121, 128
  forward-backward, 9, 16
  indirect uniqueness, 107
  induction, 108, 115
    generalized, 112
  max/min, 126, 129
  specialization, 59
  uniqueness, 105, 114
    direct, 105, 114
    indirect, 107, 114
model proof, 48
nested quantifiers, 69, 74
no (not), 81, 94, 99
NOT A implies *NOT B*, 30
NOT A, 29, 99
NOT B implies *NOT A*, 30
not of a
  quantifier, 100
  statement, 29, 99, 102
only if, 30
OR, 24, 100
overlapping notation, 26, 29
previous knowledge, 28
proof techniques
  specialization, 60
proof, 3
  by cases, 122, 129
  by contradiction, 80, 86
  by elimination, 121, 128
  condensed, 3
  model, 48
  reading, 13, 15
  techniques (see methods), 1
proposition, 29
Q.E.D., 14
quantifier, 35

choose method (for all), 48
construction method (there is), 37
existential, 35
hidden, 37, 47
nested, 69, 74
NOT, 100
specialization method, 59
universal, 35
reading proofs, 13, 15
set, 45
  defining property, 46
  element, 45
  empty, 46
  equality, 47
  member, 45
  set-builder notation, 46
  subset, 47
something happens, 36, 47
specialization method, 59–60
standard form
  for all, 47
  there is, 36
statement, 2
  conditional, 2
  contrapositive, 30
  converse, 30
  implication, 3
  inverse, 30
  not, 99, 102
subset, 47
symbolic notation, 3, 198
techniques (see methods), 1
theorem, 29
there is ("there exists" "there are"), 35
  standard form, 36
truth table, 4
  A implies *B*, 4
  NOT B implies *NOT A*, 30
uniqueness methods, 105
  direct, 105, 114
  indirect, 107, 114
universal quantifier, 35
  standard form, 47
upper bound, 60
using
  previous knowledge, 28
  the induction hypothesis, 110
working
  backward, 10
  forward, 12